Feminist Strategies in International Governance

The struggle for women's rights and to overcome gender oppression has long engaged the efforts of intergovernmental and nongovernmental organizations. *Feminist Strategies in International Governance* provides a new introduction to the contemporary forms of this struggle. It brings together the voices of academics and practitioners to reflect in particular on the effectiveness of human rights strategies and gender mainstreaming. It covers three international issue areas in which feminists currently seek change: women's human rights and violence against women; the participation of women in peace-making and their protection during conflict; and the gendered effects of development, economic and financial governance.

The book combines a critical reflection on the current state of feminist politics with an introduction to urgent issues on the contemporary international agenda. In addition, the book draws on innovative conceptualizations from constructivism in international relations, legal anthropology and discourse theory to provide new framings of current feminist struggles.

Offering an accessible guide to the engendering of international governance and examining the challenges for international feminist politics in the future, this work will be of great interest to students and scholars of international organizations, gender politics and global governance.

Gülay Caglar is a research fellow in the Division of "Gender and Globalization" at the Humboldt-University of Berlin, Germany. Her research focuses on gender politics in global economic governance.

Elisabeth Prügl is Professor of International Relations at the Graduate Institute of International and Development Studies in Geneva. Her research focuses on gender politics and international organizations.

Susanne Zwingel is Associate Professor of Politics at the State University of New York in Potsdam, NY. Her research focuses on human rights and gender in global governance.

Routledge Global Institutions Series

Edited by Thomas G. Weiss
The CUNY Graduate Center, New York, USA
and Rorden Wilkinson
University of Manchester, UK

About the series

The Global Institutions Series has two "streams." Those with blue covers offer comprehensive, accessible, and informative guides to the history, structure, and activities of key international organizations, and introductions to topics of key importance in contemporary global governance. Recognized experts use a similar structure to address the general purpose and rationale for specific organizations along with historical developments, membership, structure, decision-making procedures, key functions, and an annotated bibliography and guide to electronic sources. Those with red covers consist of research monographs and edited collections that advance knowledge about one aspect of global governance; they reflect a wide variety of intellectual orientations, theoretical persuasions, and methodological approaches. Together the two streams provide a coherent and complementary portrait of the problems, prospects, and possibilities confronting global institutions today.

The most recent related titles in the series include:

The Organization of American States (2011)
by Monica Herz

African Economic Institutions (2010)
by Kwame Akonor

Institutions of the Global South (2009)
by Jacqueline Anne Braveboy-Wagner

Institutions of the Asia-Pacific (2009)
by Mark Beeson

The African Union (2008)
by Samuel M. Makinda and F. Wafula Okumu

The European Union (2008)
by Clive Archer

Feminist Strategies in International Governance

Edited by
Gülay Caglar, Elisabeth Prügl and
Susanne Zwingel

Routledge
Taylor & Francis Group
LONDON AND NEW YORK

First published 2013
by Routledge
2 Park Square, Milton Park, Abingdon, Oxon OX14 4RN

Simultaneously published in the USA and Canada
by Routledge
711 Third Avenue, New York, NY 10017

Routledge is an imprint of the Taylor & Francis Group, an informa business

British Library Cataloguing in Publication Data
A catalogue record for this book is available from the British Library

Library of Congress Cataloging in Publication Data
Feminist strategies in international governance / edited by Gülay Caglar, Elisabeth Prügl & Susanne Zwingel.
p. cm. – (Routledge global institutions series; 70)
Includes bibliographical references and index.
1. International cooperation. 2. Women's rights–International cooperation. 3. Feminism–International cooperation. 4. Women–Political activity. 5. Women and peace. I. Caglar, Gülay. II. Prügl, Elisabeth. III. Zwingel, Susanne, 1968–
JZ1253.2.F48 2012
323.3'4–dc23
2012011990

ISBN: 978-0-415-50905-3 (hbk)
ISBN: 978-0-203-09496-9 (ebk)

Typeset in Times New Roman
by Taylor & Francis Books

MIX
Paper from responsible sources
FSC® C004839

Printed and bound by CPI Group (UK) Ltd, Croydon, CR0 4YY

Contents

Tables

Contributors

Rosa Alayza is a Sociologist, who has studied at the master and doctoral levels in Political Science. She currently teaches in the undergraduate and master programs in Political Science and Government at the Pontífica Universidad Católica in Lima (Peru). Her areas of interest are the links between state and society, public space under weak democracy and globalization conditions.

Kate Bedford is a Senior Lecturer in Law at the University of Kent. She holds a PhD in Political Science from Rutgers. Her research focuses on the interactions between sexuality studies, development studies, and international political economy. Her publications include *Developing Partnerships: Gender, Sexuality, and the Reformed World Bank* (University of Minnesota Press, 2010). She is currently researching the feminist political economy of gambling liberalization, with a particular focus on bingo.

Gülay Caglar is a Research Fellow in the Division of "Gender and Globalization" at the Humboldt-University of Berlin, Germany. She received her doctoral-level degree in Political Science at the University of Kassel. Her research focuses on modes of knowledge production and gender politics in global economic governance. She has published a book and several articles in this area, including "Multiple Meanings of Gender Budgeting: Gender Knowledge and Economic in the World Bank and UNDP" in the book *Gender Knowledge and Knowledge Networks in International Political Economy* (Nomos, 2010) edited by Brigitte Young and Christoph Scherrer.

Hilary Charlesworth is an Australian Research Council Laureate Fellow and Director of the Centre for International Governance and Justice at the Australian National University. Her research interests are in international law and international human rights law, particularly the application of feminist theories in these areas.

Carolyn Hannan is currently Associate Professor in the Human Geography Department, University of Lund, in Sweden. She worked in the United Nations Secretariat in New York for 10 years, as the Principal Officer for Gender Mainstreaming (1999–2001) and as the Director of Division for the Advancement of Women (2001–9). Prior to joining the United Nations, she worked with the Swedish International Development Cooperation Agency (SIDA) for 15 years, including both in the field and as the Senior Policy Advisor on Gender Equality in Stockholm in the 1990s. During this period, she was also part of a Swedish advisory expert group on gender mainstreaming attached to the Swedish Ministry for Gender Equality. From 1995 to 1997 she was the Chairperson of the OECD/DAC Gender Equality Network.

Anne Jenichen holds a doctoral degree in Political Science from the University of Bremen. She currently works as a senior lecturer and researcher at the Centre for European Studies of the University of Bremen. Her main research interests include norm diffusion in post-war contexts, women's rights, and religion and politics.

Jutta Joachim is Associate Professor of Political Science at the Leibniz University Hannover, Germany. She received her PhD from the University of Wisconsin-Madison and her MA from the University of South Carolina. She is the author of *Agenda Setting, the UN, and NGOs: Gender Violence and Reproductive Rights* (Georgetown University Press) and co-editor of *International Organizations and Implementation: Enforcers, Managers, Authorities* and *Transnational Activism in the UN and the EU: A Comparative Study* (both Routledge Press). Her current projects examine the role of private actors in security governance. Her articles have appeared in among others *International Studies Quarterly, German Journal of International Relations, Millennium*, and *the Journal of European Public Policy.*

Peggy Levitt is Professor of Sociology at Wellesley College and a Research Fellow at Harvard University where she co-directs the Transnational Studies Initiative. She is currently the Visiting International Fellow at the Vrije University in Amsterdam. She taught at the University of Malmö, University of Limerick, and the University of Bologna. Her books include *God Needs No Passport: Immigrants and the Changing American Religious Landscape* (New Press, 2007), *The Transnational Studies Reader* (Routledge, 2007), *The Changing*

Face of Home: The Transnational Lives of the Second Generation (Russell Sage, 2002), and *The Transnational Villagers* (UC Press, 2001). She has also edited special volumes of *International Migration Review*, *Global Networks*, *Mobilities*, and the *Journal of Ethnic and Migration Studies*. A film based on her work, *Art Across Borders*, came out in 2009.

Sally Engle Merry is Professor of Anthropology at New York University. She is the author of *Human Rights and Gender Violence* (Chicago, 2006) and *Gender Violence: A Cultural Perspective* (Wiley-Blackwell, 2009). She is president of the American Ethnological Society.

Mercedes Crisóstomo Meza holds a bachelor's degree in Anthropology from the Universidad Nacional del Centro del Peru and a magister in Political Science from the Pontífica Universidad Católica del Peru. Her research and publications are related to gender, human rights, transitional justice and memories studies. She now works as a professor in Political Science at the Pontífica Universidad Católica del Peru.

Laura Parisi is Associate Professor of Women's Studies at the University of Victoria, Canada. Her articles have appeared in journals such as *Politics and Gender*, *Journal of Human Rights*, and *Canadian Foreign Policy*. Her current research focuses on the gender mainstreaming practices of human rights and development nongovernmental organizations. Her most recent publications include: "Reclaiming Spaces of Resistance: Women's Human Rights and Global Restructuring" in *Gender and Global Restructuring: Sightings, Sites and Resistances*, 2nd ed., ed. Marianne H. Marchand and Anne Sisson Runyan (Routledge, 2011) and "Transnational" in *Rethinking Women's and Gender Studies*, edited by Catherine M. Orr, Ann Braithwaite, and Diane Lichtenstein (Routledge, 2011).

Elisabeth Prügl is Professor of International Relations at the Graduate Institute of International and Development Studies in Geneva, where she directs the Institute's Programme on Gender and Global Change (PGGC). Her research focuses on gender politics in global governance. She has published on the regulation of women's work in the International Labour Organization and the European Union, on gender mainstreaming, and on feminist theory in International Relations. Her current research probes the relationship between feminist knowledge and gender expertise.

Shahra Razavi is Senior Researcher at the United Nations Research Institute for Social Development (UNRISD). She obtained her Doctorate (DPhil) from the University of Oxford in 1993. At UNRISD she oversees the Institute's research program on Gender and Development. Shahra specializes in the gender dimensions of social development, with a particular focus on livelihoods and social policy. Her recent publications include a special issue of the *Development and Change, Seen, Heard and Counted: Rethinking Care in a Development Context* (Vol. 42, No. 4, July 2011), a co-edited (with Silke Staab) special issue of *International Labour Review*, Underpaid and Overworked: A Cross-National Perspective on Care Workers (Vol. 149, No. 4, 2010), and *The Gendered Impacts of Liberalization: Towards "Embedded Liberalism"?* (Routledge, 2009).

Rita A. Sabat is an Assistant Professor of International Relations in the Faculty of Political Science, Public Administration, and Diplomacy at Notre Dame University – Louaize, Lebanon, where she teaches courses in the areas of international institutions, gender, civil society, and development. She received her PhD in International Relations from Florida International University in Miami in May 2010. Her current research interests and specialization are in the area of conflict transformation, human rights law, gender studies, and development in the Middle East.

Andrea Schneiker studied Political Science and Sociology in Lille (France) and Münster (Germany) from 2000 to 2004. She received a PhD in Political Science from the University in Münster in 2008 for her research on the self- and co-regulation of private military and security companies. She is currently Assistant Professor at the Institute of Political Science of Leibniz University Hannover (Germany).

Jacqui True is Professor of Politics and International Relations in the School of Political and Social Inquiry at Monash University, Melbourne, Australia. Her current research interests are gender mainstreaming and global governance, gender quotas in economic decision-making, and feminist methodologies for international relations. She is the author of *The Political Economy of Violence Against Women* (Oxford University Press, forthcoming) and, with Brooke Ackerly, *Doing Feminist Research in Political and Social Science* (Palgrave, 2010).

Claudia von Braunmühl is an Honorary Professor in International Relations at the Free University of Berlin. She also has taught at the universities of Frankfurt, Edinburgh, Bielefeld and Vienna. From 1980 to 1984, she was country director of the German Development Service in Jamaica and has since worked as a consultant and advisor in development. She serves on the academic boards of Attac Germany, the Green Academy, and the Heinrich Boell-Foundation.

Mariama Williams is Senior Research Fellow with the Global Governance for Development Programme of the South Centre, an intergovernmental think tank for developing countries based in Geneva, Switzerland. She is the author of *Gender Mainstreaming in the Multilateral Trading System*; *Trading Stories: Experiences with Gender and Trade* (co-edited with Marilyn Carr); and *Gender and Trade Action Guide: A Training Resource* (with Atthill *et al.*). Mariama is a member of the International Working Group on Gender, Macroeconomic and International Economics (GEM-IWG), a Board member-at-large of the Institute of Law and Economics (Jamaica) and a member of the Board of Trustees, The Dag Hammarskjöld Foundation. She was also the former Research Coordinator with the International Gender and Trade Network, Visiting Research Economist at the Center of Concern and a Member of the DAWN Steering Committee.

Brigitte Young retired as Professor of International Political Economy from the University of Münster, Germany, in 2011. She continues to be active as a guest professor and as policy advisor on finance/Eurozone crisis. She has taught and done research at Georgetown University, Wesleyan University, CERI, Science Politique in Paris, Science-Po, Lille, the University of Warwick, and also at the Central European University, Budapest. Her research areas include economic globalization; global governance; world trade and global financial market governance; European Union-United States financial regulatory frameworks; theories of International Political Economy, and feminist macroeconomics. Her recent publications include *Questioning Financial Governance from a Feminist Perspective* (with Isabella Bakker and Diane Elson, Routledge, 2011), and several articles on the European sovereign debt crisis, and the role of Germany and France in resolving the Euro crisis.

Susanne Zwingel is Associate Professor at the State University of New York in Potsdam, United States. Her research areas include the translation of international norms, in particular women's rights norms, gender in global governance, and gender relations and armed conflict. Her book *Translating International Women's Rights: The CEDAW Convention in Context* is forthcoming from Palgrave Macmillan.

Foreword

Gülay Caglar, Elisabeth Prügl, and Susanne Zwingel's book—*Feminist Strategies in International Governance*—is the tenth in a growing number of research volumes in our "global institutions" series examining crucial global problems as well as policies and solutions to address them. These volumes serve as lengthier and more specialized treatments of given topics than is possible in the general series. As such, they are essential components in advancing the overarching aim of the series to render more visible the often complex and poorly understood world of global governance.

In addition to these longer research volumes, the series strives to provide readers with user-friendly and short (usually 50,000 words) but definitive guides to the most visible aspects of what we know as "global governance" as well as authoritative accounts of the issues and debates in which they are embroiled. We now have 70 books that act as key reference points to the most significant global institutions and the evolution of the issues that they face. Our intention has always been to provide one-stop guides for all readers—students (both undergraduate and postgraduate), interested negotiators, diplomats, practitioners from nongovernmental and intergovernmental organizations, and interested parties alike—seeking information about most prominent institutional aspects of global governance.

Caglar, Prügl, and Zwingel break new ground in the study of international organization. *Feminist Strategies in International Governance* brings together some of the very best authors working on gender and global governance critically to interrogate the gains that have been made in intergovernmental settings in the last half century. They ask: How has the recognition of gender equality as a global goal become possible? Have real gains been made which go beyond the rhetoric of gender equality? And has a sensitivity to gender really "broken out," and how profound are the transformations it has triggered? They focus on feminist strategies that have sought to normalize a sensitivity to gender differentiation and embed it in international law as well as those that pursue gender mainstreaming across international organizations. As the editors argue: "[international] organizations have played a catalytic role in focusing feminist strategizing internationally, and they have been instrumental in advancing the status of women worldwide. In turn, feminist strategizing … [has been] crucial to bringing the issue of gender equality to the UN. We suggest that, together, the UN and feminist activists have formed a unique apparatus of international governance that has made possible remarkable changes in gender regimes since the mid-twentieth century."

This is a rare and insightful book edited by three first-rate scholars. Ideally, this and other volumes in the research stream will be used as complementary readings in courses in which other specific titles in this series are pertinent—a selection of which can be found in the "About the Series" section at the front of this book. Our aim is to enable topics of importance to be dealt with exhaustively by specialists as well as enabling collected works to address issues in ways that bring more than the sum of the individual parts, while at the same time maintaining the quality of the series.

As always, we look forward to comments from our readers.

Thomas G. Weiss, The CUNY Graduate Center, New York, USA
Rorden Wilkinson, University of Manchester, UK
April 2012

Preface

A new United Nations (UN) agency, UN Women, became operational in January of 2011. Those who lobbied for the new "entity for gender equality and the empowerment of women" (as the organization is subtitled) considered its formation a historic accomplishment in a context where commitment to gender equality appeared to have hit institutional fatigue. Activists and advocates envisioned that UN Women would catalyze the UN to jump-start again the movement towards its stated goal of achieving gender equality. Hopes were running high that under the leadership of Michelle Bachelet, the charismatic former president of Chile, the organization would advance the implementation of women's rights, set up operations to support governments in achieving this goal, and ensure that entities throughout the UN system implement gender mainstreaming and consider the differential effects of their policies on women and men.

The editors of this volume took the establishment of UN Women as an occasion for a reflection on what feminists have accomplished through their work in and with international organizations. We asked ourselves, what can scholars contribute to supporting the new international momentum in favor of gender equality? What does research tell us about what has worked and what has not in international feminist strategizing? Our point of departure—and the point of departure for the contributors to this volume—is a strong feminist commitment and a belief that international strategies can serve to advance feminist agendas. At the same time, we knew that there are plenty of pitfalls in such strategies, from the co-optation of feminist agendas to the sidelining of radical demands.

Our reflection took the form of an international conference that we convened at the Graduate Institute in Geneva in October of 2010, and which brought together academics from different continents, "femocrats" in international organizations and activists. In two days of panels and roundtables we debated a range of issues including how to ensure the participation of civil society in the work of UN Women, the generational shift in women's movements with an internationalist focus, the silencing of difference in gender mainstreaming and in liberal women's rights discourse, the translation of international norms into local contexts, the failure of gender mainstreaming in the financial sector as revealed in the financial crisis, and the effects of its apparent success in certain pockets of security governance. Many of the ideas of the conference are captured in the following chapters.

We would like to thank the authors of the papers for responding to our invitation to participate in the conference and in this book project and for patiently enduring our various demands throughout the project. Their professionalism and creativity was what fueled our conversations. We owe gratitude to those who helped us organize the Geneva conference, in particular Nell Williams, who in addition to working on her own dissertation did not tire to deal with organizational details in her utterly perfectionist way. Financial support for the conference was generously provided by the Thyssen Foundation of Germany and by the Swiss National Science Foundation; we thank them for recognizing the significance of the international struggle for gender equality and for supporting our project in a thoroughly unbureaucratic fashion. In pulling together the final manuscript we received help from Hayley Thompson, another PhD candidate at the Graduate Institute; Hayley not only made sure that the manuscript's formatting met the rules, but also expertly produced the index of the book. We would like to express our deep appreciation.

We hope that readers will find this book a useful resource providing an overview over current hot topics in the engagement of feminists with international organizations. We also hope that the theories and concepts introduced and the stories told will inspire feminist activism and research. We firmly believe that the two hang together and that, in the words of one participant at the 2010 conference, "nothing is as practical as a good theory."

Gülay Caglar, Humboldt-University of Berlin
Elisabeth Prügl, The Graduate Institute, Geneva (IHEID)
Susanne Zwingel, State University of New York, Potsdam

Abbreviations

AAWORD	Association of African Women for Research and Development
ACP	African, Caribbean and Pacific Group of States
Alt-WID	Alternative Women in Development
APEC	Asia-Pacific Economic Cooperation
AWID	Association for Women's Rights in Development
BCBS	Basel Committee on Banking Supervision
BiH	Bosnia and Herzegovina
BPA	Beijing Platform for Action
CAFRA	Caribbean Association for Feminist Research and Action
CCTs	Conditional cash transfer schemes
CEAS	Episcopal Commission of Social Action (Peru)
CEDAW	Convention on the Elimination of All Forms of Discrimination Against Women
CEEWA	Council for the Economic Empowerment of Women in Africa (Kenya)
CIDA	Canadian International Development Agency
CIIR	Catholic Institute for International Relations
CLADEM	The Latin American and Caribbean Committee for the Defense of Women's Rights
CNM	National Women's Council (Argentina)
CONAMOVIDES	La Confederación Nacional de Mujeres por la Vida y el Desarrollo Integral (Peru)
CRA	Credit rating agencies
CSW	Commission on the Status of Women
DAC	Development Assistance Committee
DAW	Division for the Advancement of Women
DAWN	Development Alternatives with Women for a New Era

DCAF	Center for the Democratic Control of Armed Forces
DDR	Disarmament, demobilization, and reintegration
Demunas	Municipal Defender's Offices (Peru)
DEMUS	El Estudio para la Defensa de los Derechos de las Mujeres (Peru)
DfID	Department for International Development
DPKO	Department of Peacekeeping Operations
DRC	Democratic Republic of the Congo
EBA	European Banking Authority
ECA	European Economic Commission for Africa
ECE	Economic Commission for Europe
ECEC	Early childhood education and care services
ECLAC	Economic Commission for Latin America and the Caribbean
ECOSOC	Economic and Social Council
ECSWA	Economic and Social Commission for Western Asia
EI	Education International
EIOPA	European Insurance and Occupational Pensions Authority
EPAs	Economic Partnership Agreements
ESAs	European Supervisory Authorities
ESCAP	Economic and Social Commission for Asia and the Pacific
ESMA	European Securities and Markets Authority
ESRB	European Systemic Risk Board
EU	European Union
FAO	Food and Agriculture Organization
FATF	Financial Action Task Force on Money Laundering
FCO	Foreign and Commonwealth Office
FETs	Female Engagement Teams
FSF	Forum for Financial Stability
FTAA	Free Trade Area of the Americas
GAD	Gender and development
GATS	General Agreement on Trade in Services
GDI	Gender-related Development Index
GEAR	Gender Equality Architecture Reform Campaign
GEM-IWG	International Working Group on Gender, Macroeconomics and International Economics
GM	Gender mainstreaming
GTIA	Gender Trade Impact Assessment
HDIA	Human Development Impact Assessment

HMG	Her Majesty's Government
IA	International Alert
IAIS	International Association of Insurance Supervisors
IASB	International Accounting Standards Board
ICCPR	International Covenant on Civil and Political Rights
ICDA	International Coalition for Development Action (Brussels)
ICESCR	International Covenant on Economic, Social and Cultural Rights
ICFTU	International Confederation of Free Trade Unions
IDEA	International Institute for Democracy and Electoral Assistance
IFIs	International financial institutions
IFRS	International Financial Reporting Standards
IGOs	Intergovernmental organizations
IGTN	International Gender and Trade Network
IIF	International Institute of Finance
ILO	International Labour Organization
IMF	International Monetary Fund
INSTRAW	International Research and Training Institute for the Advancement of Women
IOs	International organizations
IOSCO	International Organization of Securities Commissions
IR	International Relations
IWGGT	Informal Working Group on Gender and Trade
IWRAW AP	International Women's Rights Action Watch Asia Pacific
LCRVAW	Lebanese Council to Resist Violence Against Women
LWN	Lebanese Women's Network
MDGs	Millennium Development Goals
MDG3	Millennium Development Goal on gender equality and empowerment of women
MDG5	Millennium Development Goal on maternal mortality
MoD	Ministry of Defence
MONUC	UN Mission in the Democratic Republic of Congo
NAFTA	North American Free Trade Area
NAPs	National action plans
NATO	North Atlantic Treaty Organization

NGOs	Nongovernmental organizations
ODA	Overseas development aid
OECD	Organisation for Economic Co-operation and Development
OHCHR	Office of the High Commissioner for Human Rights
OHR	Office of the High Representative
OP	Optional Protocol
OSAGI	Office of the UN Secretary-General's Special Advisor on Gender Issues
OSCE	Organization for Security and Co-operation in Europe
PRIO	Peace Research Institute Oslo
PROFAM	Family Strengthening and Social Capital Promotion Project
PROMUDEH	Ministry for the Promotion of Women and Human Development (Peru)
PRONAA	Programa Nacional de Apoyo Alimentario (Peru)
PWC	Post-Washington Consensus
RHRC	Reproductive Health Response in Conflict
SC	Security Council
SCR 1325	Security Council Resolution 1325
SEA	Servicios Educativos El Agustino (Peru)
Sedesol	Ministry of Social Development (Mexico)
SEWA	Self Employed Women's Association.
SIAs	Social Impact Assessments
SMART	Specific, Measurable, Achievable, Relevant and Time-bound
TPRD	Trade Policy Review Division (WTO)
TRIPS	Trade Related Aspects of Intellectual Property Rights
UK	United Kingdom of Great Britain and Northern Ireland
UN	United Nations
UNAMA	United Nations Assistance Mission in Afghanistan
UNCTAD	United Nations Conference on Trade and Development
UNDAW	see DAW
UNDP	United Nations Development Programme
UN-FfD	United Nations Conference on Financing for Development
UNFPA	United Nations Population Fund

UN-IANGE	United Nations Inter-Agency Network on Gender Equality
UNICEF	United Nations Children's Fund
UNIFEM	United Nations Development Fund for Women
UNMIK	United Nations Interim Administration Mission in Kosovo
UNRISD	United Nations Research Institute for Social Development
UNSCR	United Nations Security Council Resolutions
UNTAC	United Nations Transitional Authority in Cambodia
UNTAET	United Nations Transitional Administration in East Timor
UN Women	United Nations Entity for Gender Equality and the Empowerment of Women
USAID	United States Agency for International Development
WC	Washington Consensus
WEDO	Women's Environment and Development Organization
WEF	World Economic Forum
WICEJ	Women's International Coalition for Economic Justice
WIDE	Women in Development Europe
WTO	World Trade Organization
WWP	Women Waging Peace
WWW	Women Working Worldwide

Introducing feminist strategies in international governance

Gülay Caglar, Elisabeth Prügl and Susanne Zwingel

- **Key concepts**
- **A history of feminist encounters with multilateralism**
- **What do we know from existing literature?**
- **Organization of the book**

When the United Nations (UN) declared 1975 International Women's Year and subsequently proclaimed a Decade for Women (1976–85), feminist movements were energized worldwide and hopes were high for ending gender discrimination and women's subordination around the globe. Decades of work of women's rights advocates within the UN to create international recognition of the importance of gender equality finally seemed to bear fruit. Almost 40 years later, inequality is still pervasive, yet considerations of gender have entered the mainstream of policy-making to a degree previously unimagined. Governments around the world feel compelled to declare their commitment to gender equality—even if their actual practices often do not seem to be consistent with that goal. National and international bureaucracies in issue areas ranging from development to security proclaim their adherence to gender mainstreaming. Political parties and sometimes even private sector businesses are adopting quotas geared towards achieving gender balance. In July 2010, the UN General Assembly created UN Women, the UN agency tasked to advance gender equality and women's empowerment, which can be considered the much-delayed culmination of international feminist activism. Such activism has taken many forms and reaches back at least into the nineteenth century.[1] Since the second half of the twentieth century, it has gravitated around influencing the policy-making of international organizations with a particular focus on the UN and its sub-entities.

From the vantage point of early twentieth-century feminist activists many present-day achievements—among them, women's suffrage, the rights of working women, reproductive rights and freedom from colonial oppression—would no doubt be astonishing. How has this trend to recognize gender equality as a global goal become possible? And, those critical feminists no doubt would ask, is it really all that meets the eye? Has feminism really "broken out" around the world, and how profound are the transformations it has triggered?

In this book we approach these questions through a reflection on the concept of feminist strategies. We explore these strategies in the context of the work of intergovernmental organizations, in particular the UN and its affiliated agencies. These organizations have played a catalytic role in focusing feminist strategizing internationally, and they have been instrumental in advancing the status of women worldwide. In turn, feminist strategizing was crucial to bringing the issue of gender equality to the UN. We suggest that, together, the UN and feminist activists have formed a unique apparatus of international governance that has made possible remarkable changes in gender regimes since the mid-twentieth century. We intend for our analysis to inform the work of UN Women to help push these changes futher.

As will be laid out in more detail below, we consider international governance a realm not only entailing intergovernmental arenas, but also social and political change in domestic contexts influenced by international discourses and policies. Accordingly, the essays in this book center around feminist strategies that have aimed at influencing the inner workings of international organizations as well as their policy creation and output. Such strategies have been manifold, and we do not claim to present them all; instead, we focus on what we consider the most important contemporary multilateral feminist strategies. These can be roughly divided into two categories, namely legal or normative strategies and gender mainstreaming. Both follow an integrationist rationale: Legal or normative strategies have tried to make women's rights and gender equality norms persuasive in purportedly gender-neutral international discourses, that is, in discourses framed as technical and issue oriented, and therefore presumably not relevant to advancing gender equality. Gender mainstreaming builds on this strategy: it is a policy management tool for organizations to ensure that all its activities affect men and women equally—or put differently, that gender equality norms become operative. Thus, the objective is to include a gender perspective by systematically analyzing and assessing the gendered effects of policies, projects and programs.[2] In terms of implementing gender equality norms, gender mainstreaming is both far reaching and at risk of being co-opted within organizations where feminist knowledge or gender expertise is not widespread. The contributions in this book focus on empirical analysis, that is, they introduce concrete cases of feminist strategy creation and application in the issue areas of human rights, security and economic governance, and they evaluate the degree to which these interventions have produced change or become diluted vis-à-vis other normative or organizational claims.[3]

In this introductory chapter, we seek to provide the historical and conceptual background we consider necessary for an understanding of contemporary feminist strategizing in the multilateral arena. We first offer a definition of our key concepts, feminist strategies and international governance. This will be followed by a recounting of the major historical markers in the encounter between feminist movements and the multilateral system, including the League of Nations and the UN. We then provide an overview of contemporary literature exploring gender politics in international organizations and global governance, with an eye towards relating this literature to the issue of feminist strategizing. Finally, we offer a preview of the chapters in this book, placing them in the context of the broader literature.

Key concepts

Two key concepts guide our investigation into the politics of gender and international organizations: feminist strategies and international governance. The concepts combine highly contested terms and thus need further clarification.

Feminist strategies

The idea of feminist strategies is deceptively simple: it imagines calculating actors that target their activities so that they can bring about desired ends. When we talk about social movements or politics in general, we often conjure up precisely this image of a strategic actor. We imagine that such actors know exactly what they want, form coalitions, and engage in lobbying and advocacy until they have accomplished their ends or decide they have failed. Typically, we consider the objective of such activity the adoption of a law or, in international politics, of an international treaty or certain language in a common declaration or outcome document.

But politics and strategizing are much more complex and contingent in practice. The first problem arises with the notion of feminist actors—i.e., the "feminist" in feminist strategies. Feminist actors come in a variety of shapes, including nongovernmental organizations (NGOs), social movements, "femocrats" or specified gender units within state or international bureaucracies, and individuals such as gender experts. Organizations, specialized units and movements are of course made up of individuals, but the goals of individuals rarely coincide entirely with those of the organizations, units and movements in which they participate. While organizations and units often have mission statements and articulate their goals, the same cannot be said of movements. Indeed, movements are not unitary actors but aggregates of various organizations and individuals that agree on relatively vaguely formulated ideas and goals. Accordingly, many feminists do not speak of a single feminist movement, but of multiple movements.[4] In general, there is considerable contention as to the goal of feminism. On a principal level, the goals vary depending on the type of feminism actors make reference to. They range from achieving equality with men (liberal feminism), valorizing feminine attributes and women's social role (difference feminism), ending patriarchal subordination, oppression and exploitation such as oppressive gender divisions of labor and degrading depictions of women in the media (radical and socialist feminism), addressing the connection between gender and other status distinctions, among them race and nationality, and thus pointing to the diverse positions that women inhabit within complex social hierarchies (intersectional or transnational feminism), to entirely overcoming gender as a category that imposes stratifying order on societies (postmodern feminism).

Disagreement also exists as to how to reach feminist goals. Some feminists refer to the state as an important arena of feminist action.[5] Proponents of "state feminism"[6] regard the state or intergovernmental organizations as women friendly and attentive to feminist demands in principle. Therefore, they emphasize the potential to attain change from within; they consider the establishment of specialized gender units within bureaucracies as an essential step towards achieving gender equality and deem them important allies for women's movements.[7] In contrast, another group of feminists considers the state as patriarchal and fears the co-optation of feminist agendas. These feminists emphasize the overriding importance for women's movements to remain autonomous from the state and intergovernmental institutions.[8]

Thus, the idea of a unitary, strategizing, feminist actor is problematic. It seems that such an actor is profoundly unstable and always mobile—it is an "agent" with multiple identities working on behalf of different goals rather than an "actor" with fixed preference orderings. It takes considerable effort for such feminist agents to come together around specific goals, and agreements around such goals often are temporary, held together by a specific campaign goal only to be dissolved again once a campaign has either achieved its goal or fizzled out.[9] Thus, when we talk about feminist strategies in the plural, we have in mind not only different types of strategies, but also different types of agents and goals.

A second problem with a rationalist notion of feminist strategizing has to do with the assumption that political processes proceed in a straightforward fashion. There clearly is some relationship between the means and ends of feminist strategizing, but there are a range of factors that complicate this relationship. For one, as already mentioned, goals are not fixed in meaning. Rather, they are constantly being negotiated and adjusted internally and depending on negotiations with partners and with allies in international bureaucracies. Furthermore, organizational politics intervene to deflect or detour the intentions of feminist strategies. The case of gender mainstreaming shows that when bureaucracies officially adopt a policy, implementation is not guaranteed. The policy may encounter resistance, may be co-opted to different ends or subverted to become meaningless.[10] Finally, national contexts matter in the way in which a policy or law is being implemented. For example, women's rights codified in international treaties are being interpreted in local contexts and matched to fit local practices.[11] In other words, strategies develop their own logic and acquire new meanings once they are deployed in different organizational and socio-political contexts. Unleashed from the intentions of agents, they gain a life of their own.

A rationalist understanding of strategies also cannot account for a larger rationality that derives from the intent to govern. We typically think of democratic government—including international institutions—as a benign apparatus that has the best of the world's population in mind. However, it is necessary to question the practice of governing itself. In the international arena in particular, government is often highly technocratic and provides little room for democratic processes. Instead, a variety of experts propose solutions that have been identified as "global problems."[12] A significant cadre of gender experts has emerged in the capitals of the world and in New York and Geneva, advising governments and international organizations. These experts produce codes of conduct and formulate laws; they identify best practices and rank countries according to how well they do with regard to expert-defined standards of gender equality. Thus, they produce policy-relevant knowledge, which—in Foucault's words—"induces effects of power."[13] In other words, knowledge makes things governable and regulates the conduct of others by constantly drawing the boundaries of acceptable behavior in a specific policy field. While these practices amount to a kind of feminist politics, their primary effect may be to govern. Thus, in becoming part of international institutions, feminist strategizing develops its own rationality: to govern through subtle and indirect means. In this context, feminist strategizing is turned from a mode of resistance into an instrument of power.

In sum, when we talk about feminist strategies in this book, we have in mind a complex and multifaceted phenomenon of exerting influence. We are interested in how feminist goals are being negotiated, which alliances are formed, how policy problems are being framed, represented and redefined, how policy successes are translated in different political spaces, and how radical intentions are turned into a mode of government.

International governance

The expressions "international governance" or "global governance" have largely replaced the term "international organization" to designate politics that involve the multilateral arena. Part of the reason has been to signal a shift away from studying politics in organizations to studying broader institutional phenomena, including international regimes negotiated in the context of international organizations and watched over by these organizations. Another part has been to shed light on non-institutional policy processes, such as the influence of social movements that has accelerated rapidly in the second half of the twentieth century.[14] In choosing the term governance in our title, we align ourselves with literature that has highlighted these phenomena.

Our choice of the more old-fashioned term "international" over "global" needs a little more explaining. We think that the term global should be reserved to describing transnational flows and connections associated with processes of globalization, such as movements of people, goods, ideas, and money or the creation of border-defying spaces of communication. These movements and spaces are no doubt governed; but such global governance does not necessarily have its origins in political institutions, at least not exclusively and perhaps not even primarily. New global cultures, hybridized or hegemonically constituted, new forms of identity, new configurations of hierarchy govern these global, social spaces with or without intersecting pre- and proscriptions from formal political bodies.[15] The norms and structures that operate in these spaces thus encompass much more than formal rules and laws; they comprise in addition the cultures, habits, and routines that emerge from everyday interactions. As an inflection of governance, we thus take the term global to signify a plethora of border-crossing rules and norms that span the realms of politics, society, and culture.

This book recognizes governance in global spaces but does not start from there. Instead, our starting point is the intergovernmental arena, the world of international organizations and of the world cultures that they participate in creating. This is an arena in which states decidedly matter. It is also an arena where formal institutions and bureaucracies generate formal and informal rules that circumscribe a variety of activities in the contemporary world—from the way we think about violence against women to the legitimacy of denying or conceding basic labor rights. In this arena, governments may be attentive to the contributions of NGOs and of private actors. They also may find themselves confined in their agency by systemic forces such as international movements of capital. But they and the institutions they have created remain key players, targets of activism, and gatekeepers to international influences on national and local societies. Our focus of attention is this world of interstate politics including the autonomous spheres of action that international organizations have carved out for themselves, and the way in which feminist strategies have operated in this political environment.

A history of feminist encounters with multilateralism

When did feminists start to influence multilateral institutions? What were their goals and strategies? The history of feminist activism in the League of Nations is only beginning to be written, but such activism was significant. The Treaty of Versailles, which founded both the League of Nations and the International Labour Organization (ILO), included language on women requiring that positions in the League secretariat and in the ILO be open to women, that the ILO consider the protection of women, and that women be included in government delegations when topics that concerned them were discussed. This language entered the treaty largely as a result of intense feminist lobbying at Versailles. Once the League was established, feminists considered it an important arena in which to advance their agendas. In 1925, 11 feminist organizations formed the Joint Standing Committee of Women's International Organizations (later much enlarged and renamed the Liaison Committee of Women's International Organizations), which regularly met in Geneva and London and became the primary interlocutor for the League on women's issues. The Committee worked on the promotion of world peace, targeted the disarmament conferences of the late 1920s, took up the issue of women's economic rights (including in the aftermath of the great depression combating efforts to restrict women's right to work), fought trafficking and prostitution, and issued statements on a variety of other issues. Liberal feminist organizations floated the idea of an equal rights treaty causing a rift with socialist and union women who feared that such a treaty threatened hard-fought protective labor legislation.[16]

In addition to lobbying the League and the ILO from the outside, feminists also were active inside the new international organizations, where they served as experts and staff, such as on the League's Advisory Committee on Traffic in Women and Children or in the ILO's Bureau of Women and Children.[17]

After 1945, feminists resumed their internationalist activism by targeting the newly formed UN.[18] According to Peggy Antrobus, this activism can be divided into three broad phases reflecting the institutional transformations it generated: "international knowledge formation" from 1945 to 1974, "global transformation" from 1975 to 1995, and "consolidation and shift of allegiances" from 1996 to the present.[19] The first phase—international knowledge formation—has been described as a time when "the tool of language (was used) to jump-start change" within the UN in regard to the status of women.[20] The Commission on the Status of Women (CSW)—founded in 1946 and at the time both an institutional and movement-oriented women's space—was paramount in this period. Both legal and socio-economic dimensions of the inferior status of women were framed as internationally relevant. The *Declaration on the Elimination of Discrimination against Women* of 1967 can be seen as a symbol for the comprehensive understanding of women's subordination, but it did not trigger much change. The limits of a purely institutional strategy became apparent.

The second phase—global transformation—in retrospect looks almost like an explosion of international awareness raising, both in terms of the publicity that gender equality received and the multiplicity of actors involved. The 1976–85 UN Decade for Women and the corresponding conferences produced three important innovations: First, it spawned new tools and institutions to work for gender equality on all levels, including the United Nations Development Fund for Women (UNIFEM) and the Convention on the Elimination of All Forms of Discrimination Against Women (CEDAW). Second, the global dialogue made possible by the conferences enabled a learning process that first acknowledged differences among women and then envisaged joint strategies of action. A crucial factor in this learning was the articulation of women's voices from the south. The decade produced, third, a vibrant, articulate, and diverse global women's movement whose main nodes were transnational networks[21] knowledgeable about reframing global justice issues in a variety of fields and influencing various UN conferences in the 1990s—including conferences on the environment and development (1992), human rights (1993), population policy (1994), and social development (1994). At the Beijing Fourth World Conference on Women in 1995 feminist activists demanded participation of women in formal channels of power, winning support for gender mainstreaming and a comprehensive catalogue of state commitments in the Beijing Platform for Action (BPA). As Antrobus points out, however, this impressive global process was paralleled by a shift from state-oriented towards market-driven development policies that had a negative impact on the living situations of the poor, and of poor women in particular. Thus, the irony was that during a phase of high level sensitization regarding gender hierarchies worldwide, the living conditions of many women deteriorated.[22]

The last phase—consolidation and shift of allegiances, from 1996 to the present—departs from the Beijing conference, arguably the climax in terms of global awareness raising and thus probably a natural point of decline in terms of UN-focused gender activism. This phase is shaped by disenchantment, as a downward, or at best stagnating, trend in terms of women's socio-economic conditions has persisted in spite of the international and national commitment to gender equality articulated at the world women's conferences.[23] As the UN was not able to address global injustices vis-à-vis the neoliberal development framework proposed by international financial institutions—which became particularly clear in the context of the UN Conference on Financing for Development 2002 in Monterrey, Mexico—the typical movement actors from the 1980s and 1990s somewhat receded from UN-related activism. Some organizations found independent events such as the World Social Forum more adequate to develop visions of global justice; others reoriented their work to national and local levels or to implementation activities of global norms. Nonetheless, during this phase gender awareness in international institutions strengthened significantly, as expressed in attempts to engender global security policies and in the creation of UN Women.

What do we know from existing literature?

How do feminist scholars assess the achievements of feminist interventions so far, and how do they evaluate the potential of feminist strategies to trigger further change? The literature on gender politics in international governance can broadly be divided into two categories: actor-specific literature and writings that focus on certain aspects of institutional strategies in different issue areas.

The vibrant activism of transnational women's networks has drawn the attention of feminist scholars who have investigated the organizational characteristics of such networks and their tactics to bring gender issues onto the agenda of intergovernmental conferences and organizations. This literature oscillates between positive and restrained assessments of feminist interventions: On the positive side, feminist scholars stress that transnational women's networks "were very influential in problem definition and agenda setting and had some impact on policy formulations"[24]—this is considered an achievement, even if the influence on implementation processes at the national level is limited. International Relations (IR) scholars, particularly those drawing on social movements theory, show how feminist actors engage in different forms of "strategic framing"[25] or "discursive politics"[26] to legitimize certain gender equality claims. The notions of strategic framing and discursive politics suggest that feminist actors articulate problems and solutions they consider appropriate in ways that make them plausible and acceptable for policy-makers. Joachim points out that the acceptance of such frames depends on political opportunity structures and mobilization structures.[27] Political opportunity structures include changing political constellations, influential allies or symbolic events which change the political situation or the perception of a problem and consequently open avenues to influence the policy process. An example would be the World Bank's partial opening for interventions from women's NGOs in the mid-1990s. This shift was due to a change in leadership that seemed at least rhetorically committed to gender equality and to the general attempt to domesticate feminist critiques of World Bank policies.[28] Whether and to what extent feminist actors are able to seize these opportunities depends on their mobilization structures, i.e., intellectual resources, credibility, and internal capacity to mobilize for change. As many scholars show, an important source of mobilization is the degree to which feminist actors are organized in networks, i.e., in transnational advocacy networks[29] or "velvet triangles"[30] that bring together activists with different strengths and resources working out of a variety of contexts. The contribution of Caglar in this volume adds to this literature and exemplifies that knowledge-based advocacy networks particularly contribute to feminist mobilization on economic governance.

The more critical assessments of feminist interventions emphasize the conservative political environment—the neoliberal framework and different kinds of fundamentalisms—feminist actors are currently confronted with. As Mary Hawkesworth states, within this context, "outsider politics is often marginalized as utopian, while insider politics is often co-opted […] or derailed by the standard operating procedures of gendered institutions."[31] Thus, political opportunity structures for gender politics in both agenda-setting and implementation processes have closed off drastically. Feminist scholars, especially those with a postcolonial orientation, have long expressed a critical stance towards feminist organizing within a neoliberal context.[32] Their analyses show that women's organizations run the risk of being co-opted either as experts or as local service providers in the operational work of, for instance, international development agencies.

Clearly, feminist actors operating from the inside face powerful challenges: They are confronted with resistant organizational cultures; tenacious institutional routines and cognitive frameworks that counteract the integration of gender equality demands. One strategy to bring gender issues onto the agenda has been to accommodate the restrictions imposed by these structural barriers. Literature on institutions of economic governance reveals how gender issues are fitted into the neoliberal economic paradigm by using the dominant language or by referring to instrumental arguments, like—as in the case of the World Bank—the economic utility of gender equality or women's contribution to economic growth.[33] Thus, gender issues have by all means been recognized in different areas of international governance, yet, mostly at the expense of the intrinsic value of gender equality. This is lucidly illustrated by the contributions of Razavi and Bedford in this volume: they demonstrate how issues of gender equality in economic governance are subsumed under (child-centered) family policies and how, that way, intimate relations at the private level are governed.

The set of feminist literature on institutional strategies encompasses literature on international law and gender mainstreaming. Feminist strategies to engender international law, and in particular human rights law, precede feminist activism related to economic and security governance. While feminist legal strategies have faced significant resistance,[34] the issue area of human rights proved to be a promising field for the integration of gender, partly because a gender dimension was already inherent in it, but not sufficiently spelled out,[35] and partly because of the promotional character of human rights law.[36] The rhetorical breakthrough of feminist interventions in this field notwithstanding, feminist literature continuously criticizes that the integration of gender into legal frameworks has remained insufficient (see Charlesworth in this volume). And while gender norms have become surprisingly powerful in international discourses, they require forceful and continuing feminist activism to retain their transformative potential on the international level itself, as Zwingel's contribution shows. Furthermore, international gender norms need to undergo a process of contextualization when domestically implemented that pushes women's rights into the real lives of women. The characteristics of such processes, including enhancing or undermining governmental and nongovernmental actor constellations, have become the object of a growing number of transnational feminist studies (see also Levitt *et al.* and Sabat in this volume).[37]

Literature on gender mainstreaming as a theoretical concept and as a strategy which is now widely adopted in global, regional and national contexts has mushroomed in the last decade. It includes literature aimed at facilitating the implementation of gender mainstreaming in different issue areas,[38] critical analyses of the institutional opportunities for and obstacles to the implementation of gender mainstreaming,[39] and a vibrant discussion on the theoretical implications of practicing the instrument, in particular if its transformative effect on gender relations is considered.[40] Interestingly, analyses of the latter category show that gender mainstreaming policies notably vary depending on the meaning that is assigned to gender issues in general and to the goal of gender equality in particular. These meanings are not fixed but rather shaped by discursive strategies; thus, they are a matter of contestation.[41] The construction of meanings is inextricably linked to gender expertise; gender experts exercise power by deploying technologies of knowledge and by framing the problem that has to be solved in specific ways as the contribution of Prügl in this volume shows.

Organization of the book

Part I of the book provides an overview and critical scrutiny of feminist strategies applied in international governance. The authors reflect on the use of law and gender mainstreaming, the development of various forms of gender expertise and the transformative potential of such knowledge, and on insider approaches taken by bureaucrats within international organizations versus outsider strategies employed by civil society actors. Hilary Charlesworth argues that the integration of gender equality into international human rights law has only yielded very limited improvements in women's lives. She pleads for a more radical engendering of human rights that goes beyond the standard of a male comparator. Jacqui True and Laura Parisi provide a critical introduction to gender mainstreaming, the primary feminist strategy employed by international organizations today. They note continued widespread resistance, a reliance on essentialist assumptions of what it means to be a woman, and the absence of evaluation efforts in regard to gender mainstreaming practices. The authors call for reforming the ways in which the strategy is currently applied. They stress in particular that the critical "outsider" assessment of women's movements should be integrated into organizational gender mainstreaming practices. Elisabeth Prügl analyzes the strategic quality of gender expertise as revealed in gender training manuals in the field of security governance. She shows that such expertise becomes strategic for a variety of ends, that it makes use of different types of technologies that convey it authority, and that it develops different logics that generate both empowerments and disempowerments. Carolyn Hannan, former Head of the UN Division for the Advancement of Women, highlights feminist achievements within the UN context as well as difficulties in maintaining these achievements in light of recent backlashes. She emphasizes strategies of coalition building with powerful, willing actors that dispose of significant resources, alerting us to the skillful tightrope walking that has been necessary for success in a highly masculinized environment. Mariama Williams sheds light on the interventions of feminist activists in the field of international trade by re-evaluating the experiences of a global south-centered transnational feminist network—the International Gender and Trade Network (IGTN). She highlights the importance of IGTN's networking strategy and explains its critical stance towards the strategy of gender mainstreaming within the context of the World Trade Organization.

Part II of the book elaborates on international gender equality norms with a particular focus on CEDAW. All contributions stress the malleability of gender norms, both in regard to their interpretation within international contexts and in national and local spheres. Susanne Zwingel sheds light on the continuous need of the CEDAW committee to construct its legitimacy on the international level. This legitimacy depends both on intergovernmental approval and outspoken advocacy for women's rights, yet a balance between these two demands means to avoid being labeled as a committee of "radical feminists." The author makes the point that while the actual aim of the committee is norm implementation, it also needs continuous "norm work" just to stay internationally authoritative. Peggy Levitt and co-authors analyze how global ideas and practices on women's human rights are understood and used on the ground. They introduce the concept of vernacularization in order to describe how the idea of women's human rights has been modified with the purpose of making it acceptable in the Peruvian context. The chapter illustrates how international norms encounter local struggles and different political strategies of civil society actors. Rita Sabat exemplifies how the international norm of combating violence against women gets translated at local levels in Lebanon. She illustrates that international norms do not simply diffuse into local contexts but go through a process of translation which spawns a change of the pertaining norm. Sabat argues that agents strategically adapt to domestic social and political structures and make use of the context-specific repertoires of interpretation when arguing for the implementation of international norms.

Part III elucidates the engendering of the security discourse through various Security Council resolutions and then explores the extent to which these international standards have been applied domestically. Claudia von Braunmühl's analysis of attempts and pitfalls to engender the international security discourse detects that international institutions have prioritized the notion of protecting women as victims over their empowerment and participation in designing security policies. Like Charlesworth, and True and Parisi, she is critical of integrative approaches whose transformative potential she considers hemmed in by the liberal peace assumptions inherent in UN discourse that preclude more radical conceptualizations of positive peace. Andrea Schneiker and Jutta Joachim compare the implementation strategies that three Western European states (Great Britain, Germany and Sweden) have developed in order to reframe their security policies according to Security Council Resolution 1325. The authors find widely differing interpretations of the meaning of 1325 which translate into integrationist and transformative reform proposals, supporting both critical and more optimistic assessments of feminist insider strategies. Anne Jenichen sheds light on the limited space that post-war societies provide for the implementation of gender equality norms. As she shows for Bosnia and Herzegovina, most political actors, including those representing international organizations, consider these norms secondary for the building of a peaceful post-conflict society. Nevertheless, the presence of international peacebuilding operations opens a window of opportunity for domestic proponents of gender equality and allows them to form coalitions with international femocrats supportive of feminist policy change.

Part IV focuses on the field of international economic governance. The theoretical orientation in this part is less concerned with the spread of norms, but with the effects of expert knowledge and with the absence of women in governance institutions in the case of finance. Shahra Razavi surveys the way in which governments around the world have turned away from the neoliberal Washington Consensus of the 1980s to socially "re-embed" economic governance in different ways. She introduces some social programs illustrating this re-embedding; these seek to ensure against social risk and foster social investments largely by focusing on investing in children. The status of grown-up women and mothers in labor markets falls by the wayside in this new approach. Kate Bedford further develops the notion of economic governance by extending it to the "very local" level of intimate family politics. She analyzes the World Bank's programs in Latin America and shows that these programs construct a new understanding of gender roles in the family—firmly embedded within a regime of normative heterosexuality but with males called on to take responsibility while women are enabled to go out to work. Taking seriously the governance of intimacy reveals thus another form in which integrative strategies co-opt feminist agendas. Gülay Caglar asks why national and international responses to the economic crisis have ignored gendered impacts despite vibrant feminist activism in the field of economic governance and a plethora of alternative ideas on how to engender macroeconomic policies. She traces the discourse of feminist economics in international organizations (World Bank and the United Nations Development Programme) to show that there has been a pervasive refusal to consider gender a core economic issue. Brigitte Young argues that the global financial crisis of 2008 was man-made. The blatant exclusion of women from the boards of high finance and financial economics, facilitated by a shift from public to semi-private market governance, fostered a situation of groupthink, in which a homogenous group failed to receive important checks on reality. To Young, adding women to financial governance structures is a first step towards counteracting this dangerous situation.

As a group, the contributors to this volume thus weigh the benefits against the steep costs of feminist insider strategies, shed light on the constant need for persuasion and creating legitimacy for feminist demands, and highlight the fragility of achievements and small step successes. The patterns of these findings are outlined in the conclusion, which also discusses implications for future research, theory building and policy.

Notes

1 Deborah Stienstra, *Women's Movements and International Organizations* (New York: St. Martin's Press, 1994); Margaret E. Galey, "Forerunners in Women's Quest for Partnership," in *Women, Politics, and the United Nations*, ed. Anne Winslow (Westport, CT: Greenwood Publishing Group, 1995), 1–10; Leila J. Rupp, *Worlds of Women: The Making of an International Women's Movement* (Princeton, NJ: Princeton University Press, 1997).

2 Jacqui True, "Mainstreaming Gender in International Institutions," in *Gender Matters in Global Politics: A Feminist Introduction to International Relations*, ed. Laura J. Shepherd (New York: Routledge, 2010), 190; see contribution of Jacqui True and Laura Parisi in this volume.

3 We focus on human rights, security and economic governance, three fields that encompass a wide range of topics feminists currently pursue in their international strategizing. Yet other important topics fell by the wayside, including population politics, environmental politics, and the politics of HIV/AIDS. The fact that we neglect them here should not detract from their key importance to international feminist strategizing.

4 See Amrita Basu, ed., *The Challenge of Local Feminisms: Women's Movements in Global Perspective* (Boulder, CO: Westview Press, 1995).

5 See Amy G. Mazur and Dorothy E. McBride, "State Feminism since the 1980s: From Loose Notion to Operationalized Concept," *Politics & Gender* 3, no. 4 (2007): 501–13.

6 Helga Hernes defines state feminism as "a variety of public policies and organizational measures, designed partly to solve general social and economic problems, partly to respond to women's demands." Helga Hernes, *Welfare State and Woman Power: Essays in State Feminism* (Oslo: Norwegian University Press, 1987), 11.
7 Alison E. Woodward, "Building Velvet Triangles: Gender and Informal Governance," in *Informal Governance in the European Union*, ed. Thomas Christiansen and Simona Piattoni (Northampton, MA: Edward Elgar, 2003), 76–93.
8 Stienstra, *Women's Movements and International Organizations*; see Birgit Sauer, *Die Asche des Souveräns: Staat und Demokratie in der Geschlechterdebatte* (New York: Campus Verlag, 2001).
9 Margaret E. Keck and Kathryn Sikkink, *Activists Beyond Borders: Advocacy Networks in International Politics* (Ithaca, NY: Cornell University Press, 1998).
10 Elisabeth Prügl, *Transforming Masculine Rule: Agriculture and Rural Development in the European Union* (Ann Arbor, MI: University of Michigan Press, 2011).
11 Susanne Zwingel, *Translating International Women's Rights: The CEDAW Convention in Context* (New York: Palgrave, forthcoming).
12 Gülay Caglar, *Engendering der Makroökonomie und Handelspolitik: Potenziale transnationaler Wissensnetzwerke* (Wiesbaden, Germany: VS Verlag für Sozialwissenschaften, 2007).
13 Michel Foucault, "Prison Talk," in *Michel Foucault: Power/Knowledge*, ed. Colin Gordon (New York: Pantheon Books, 1980), 52.
14 Robert O. Keohane and Joseph S. Nye Jr., "Introduction," in *Governance in a Globalizing World*, ed. Joseph S. Nye and John D. Donahue (Washington, DC: Brookings Institution Press, 2000), 1–41.
15 Arjun Appadurai, *Modernity at Large: Cultural Dimensions of Globalization* (Minneapolis, MN: University of Minnesota Press, 1996).
16 Karen Garner, *Shaping a Global Women's Agenda: Women's NGOs and Global Governance, 1925–85* (Manchester: Manchester University Press, 2010); Elisabeth Prügl, *The Global Construction of Gender: Home-based Work in the Political Economy of the 20th Century* (New York: Columbia University Press, 1999).
17 Gyoung Sun Jang, "The Sexual Politics of the Interwar Era of Global Governance: Historicizing the Women's Transnational Movements with(in) the League of Nations, 1919–40," PhD diss., Clark University, 2009; Carol Riegelman Lubin and Anne Winslow, *Social Justice for Women: The International Labor Organization and Women* (Durham, NC: Duke University Press, 1990).
18 Stienstra, *Women's Movements and International Organizations*, 67–75.
19 See in more detail Peggy Antrobus, *The Global Women's Movement. Origins, Issues and Strategies* (London: Zed Books, 2004); Devaki Jain, *Women, Development, and the UN. A Sixty-Year Quest for Equality and Justice* (Bloomington, IN: Indiana University Press, 2005); Valentine M. Moghadam, *Globalizing Women. Transnational Feminist Networks* (Baltimore, MD: Johns Hopkins University Press, 2005).
20 Jain, *Women, Development, and the UN*, 7.
21 Elisabeth J. Friedman, "Women's Human Rights: The Emergence of a Movement," in *Women's Rights, Human Rights: International Feminist Perspectives*,

ed. Julie Peters and Andrea Wolper (London, New York: Routledge, 1995), 18–35; Moghadam, *Globalizing Women.*

22 Antrobus, *The Global Women's Movement*, 54.

23 Ibid.; Jain, *Women, Development, and the UN.*

24 Ilse Lenz, "Transnational Social Movement Networks and Transnational Public Spaces: Globalizing Gender Justice," in *Rethinking Transnationalism: The Meso-link of Organizations*, ed. Ludger Pries (New York: Routledge, 2008), 105.

25 Jutta Joachim, *Agenda Setting, the UN, and NGOs: Gender Violence and Reproductive Rights* (Washington, DC: Georgetown University Press, 2007).

26 Mary E. Hawkesworth, *Globalization and Feminist Activism* (Lanham, MD: Rowman and Littlefield, 2006), 80.

27 Joachim, *Agenda Setting, the UN, and NGOs.*

28 Robert O'Brien, Anne Marie Goetz, Jan Aart Scholte and Marc Williams, *Contesting Global Governance* (Cambridge: Cambridge University Press, 2000).

29 Debra J. Liebowitz, "Gendering (Trans)National Advocacy," *International Feminist Journal of Politics* 4, no. 2 (2002): 173–96.

30 That is networks of women's NGO representatives, femocrats, and gender experts. Woodward, "Building Velvet Triangles."

31 Hawkesworth, *Globalization and Feminist Activism*, 108.

32 Sonia E. Alvarez, "Advocating Feminism: The Latin American Feminist NGO 'Boom,'" *International Feminist Journal of Politics* 1, no. 2 (1999): 181–209; Sonia E. Alvarez, "Beyond NGOization? Reflections from Latin America," *Development* 52, no. 2 (2009): 175–84; Gayatri Chakravorty Spivak, "'Woman' as Theatre: United Nations Conference on Women, Beijing 1995," *Radical Philosophy* 75 (1996): 1–7; Anne Sisson Runyan, "Women in the Neoliberal Frame," in *Gender Politics in Global Governance*, ed. Mary K. Meyer and Elisabeth Prügl (Lanham, MD: Rowman & Littlefield Publishers, 1999), 210–20.

33 Shahra Razavi, "Becoming Multilingual: The Challenges of Feminist Policy Advocacy," in *Missionaries and Mandarins. Feminist Engagement with Development Institutions*, ed. Carol Miller and Shahra Razavi (London: Intermediate Technology Publications, 1998), 20–41; Nüket Kardam, *Bringing Women In: Women's Issues in International Development Programs* (Boulder, CO: Lynne Rienner Publishers, 1991); Kate Bedford, *Developing Partnerships: Gender, Sexuality and the Reformed World Bank* (Minneapolis, MN: University of Minnesota Press, 2009); Penny Griffin, *Gendering the World Bank. Neoliberalism and the Gendered Foundations of Global Governance* (New York: Palgrave Macmillan, 2009).

34 Hilary Charlesworth, "Human Rights as Men's Rights," in *Women's Rights, Human Rights: International Feminist Perspectives*, ed. Julie Peters and Andrea Wolper (New York: Routledge, 1995), 103–13; Christine M. Chinkin, "Women's Rights as Human Rights under International Law," in *Understanding Human Rights*, ed. Conor Gearty and Adam Tomkins (London: Mansell Publishing, 1996), 553–72.

35 Since the 1948 Universal Declaration of Human Rights, non-discrimination based on sex has been an explicit part of the international human rights framework.

36 Jack Donnelly, "International Human Rights: A Regime Analysis," *International Organization* 40, no. 3 (1986): 599–642. The promotional character

of the international human rights framework situates norm formulation on the international level, but norm realization in the domestic realm. Ideally, these two should be connected, but the connection is made by governments without external interference. This disconnect makes the norm creation process more open than in other fields with direct domestic repercussions—for example, international trade agreements.

37 Among others, Amina Jamal, "Transnational Feminism as Critical Practice. A Reading of Feminist Discourses in Pakistan," *Meridians: Feminism, Race, Transnationalism* 5, no. 2 (2005): 57–82; Peggy Levitt and Sally Merry, "Vernacularization on the Ground: Local Uses of Global Women's Rights in Peru, China, India and the United States," *Global Networks* 9, no. 4 (2009): 441–61; Millie Thayer, *Making Transnational Feminism. Rural Women, NGO Activists, and Northern Donors in Brazil* (New York: Routledge, 2010).

38 For example, Mariama Williams, *Gender Mainstreaming in the Multilateral Trading System. A Handbook for Policy-Makers and Other Stakeholders* (London: Commonwealth Secretariat, 2003).

39 For example, Rebecca Tiessen, *Everywhere/Nowhere: Gender Mainstreaming in Development Agencies* (Sterling, VA: Kumarian Press, 2007).

40 For example, Carol Bacchi and Joan Eveline, ed., *Mainstreaming Politics: Gendering Practices and Feminist Theory* (Adelaide: University of Adelaide Press, 2010); Judith Squires, *The New Politics of Gender Equality* (New York: Palgrave Macmillan, 2007); Mieke Verloo, ed., *Multiple Meanings of Gender Equality. A Critical Frame Analysis of Gender Policies in Europe* (Budapest: Central European University Press, 2007)

41 Emanuela Lombardo, Petra Meier and Mieke Verloo, ed., *Discursive Politics of Gender Equality* (New York: Routledge, 2009).

Part I

Feminist strategies

1 International human rights law

A portmanteau for feminist norms?

Hilary Charlesworth

- **Sex and gender in international law**
- **Women's rights after conflict**
- **Conclusion**

What is the relationship of international human rights law to sex and gender? Is international law a useful mechanism to improve women's lives? International law presents itself as a universal system of values whose breach has legal repercussions, including sanctions, triggering a duty to compensate, or sparking condemnation by international bodies. Breaching international law may have a further impact domestically if international law is incorporated into a national legal system. Violations of international law can also have political effect through reducing the legitimacy of particular courses of action or inaction.

This chapter considers, first, international human rights law's account of sex and gender and the way it has shaped rules relating to women's lives. It then explores how these rules have been disseminated and the effect they have had in the context of post-conflict societies. It argues that, while international institutions have developed a suite of legal standards relating to sex and gender, the observance of these standards has been limited. Feminist concepts and methods must be developed to give international law standards some bite. The new United Nations (UN) entity dealing with women, UN Women, is well placed to begin this task.

Sex and gender in international law

Understandings of sex and gender in international human rights law have always revolved around women, and usually the context has been that of discrimination. This means that their development has taken place in particular fora and contexts and that sex and gender are understood as a "special" interest, distinct from mainstream concerns of the discipline. Men's lives are rarely viewed through these lenses.

The major focus of the protection of women's rights internationally has been the right to equal treatment and non-discrimination on the basis of sex. The UN Charter (1945) was the first international agreement to endorse the equality of the sexes, the product of an energetic lobbying campaign by women's organizations with the support of a few women representing states at the San Francisco conference. In its preamble, the Charter reaffirmed faith in "the equal rights of men and women" and the purposes of the UN included promotion of human rights "without distinction as to race, sex, language or religion."[1] The Charter also specified that the UN should "place no restrictions on the eligibility of men and women to participate in any capacity and under conditions of equality in its principal and subsidiary organs."[2]

Women's rights were also on the agenda of the drafters of the Universal Declaration of Human Rights, adopted by the UN in 1948. The UN Commission on Human Rights was given a specific mandate to deal with the status of women in the Declaration. Signs of the influence of women delegates and institutions in drafting the Universal Declaration are the reference to the equal rights of men and women in the Preamble, Article 2's prohibition of discrimination on the basis of sex with respect to the enjoyment of rights contained in the Declaration and Article 16's guarantee of the equal rights of women and men before, during and after marriage.

The two major human rights instruments, the International Covenant on Economic, Social and Cultural Rights (ICESCR) and the International Covenant on Civil and Political Rights (ICCPR), both adopted in 1966, contain general provisions relating to women's rights. Both covenants provide that the rights they recognize should be respected "without distinction of any kind" including sex,[3] and, more specifically, that state parties should ensure the equal right of men and women to enjoy the designated rights.[4] The ICCPR also offers a broad guarantee of equality before the law and a prohibition on discrimination on a number of grounds, including sex.[5] International human rights law thus provides the bones for a body of jurisprudence on women's rights.

The most wide-ranging of the international human rights treaties devoted to women is the Convention on the Elimination of All Forms of Discrimination Against Women (CEDAW), adopted by the UN General Assembly in 1979. The Convention contains a broad definition of discrimination in Article 1, covering both equality of opportunity (formal equality) and equality of outcome (de facto or substantive equality). It requires states to take legal and other measures to ensure the practical realization of the principle of sex equality.[6] The Convention covers a range of areas where state parties must work to eliminate discrimination including political and public life, international organizations, education, employment, healthcare, financial credit, cultural life, the rural sector, and the law.[7] In its wide coverage, the Women's Convention transcends the traditional divide between civil and political rights, and economic, social and cultural rights, illustrated by the separate development of the ICESCR and the ICCPR.

The Women's Convention also offers a bridge between public and private spheres of activity. This is unusual in international law which typically operates to regulate "public" areas such as resources, or relations between states, or human rights abuses directly by the state, but leaves "private" areas such as the family unregulated. The Women's Convention, for example, explicitly affirms women's right to equality within the family.[8]

One striking omission from the Women's Convention is the prohibition of violence against women. This may be because, at the time of the treaty's adoption, the global extent of violence against women was not well understood, or because violence was not analyzed as a matter of discrimination. The Women's Convention monitoring body, the Committee on the Elimination of Discrimination Against Women, adopted a General Recommendation on Violence against Women in 1992, stating that "gender based violence is a form of discrimination which … impairs or nullifies the enjoyment by women of human rights and fundamental freedoms."[9]

The Women's Convention has attracted an impressive number of parties (187 as of 1 October 2011). Many states have made broad reservations to provisions of the treaty, however, which limit their commitment to it. Most of these reservations are based on assertions of cultural and religious values. Some states have objected to the wide reservations,[10] and CEDAW has probed reserving states on this issue. Under the international legal regime governing reservations, no more direct sanctions can be used to pressure states to withdraw their sweeping reservations.[11]

An Optional Protocol to the Women's Convention was adopted in 1999 to allow for both individual communications to CEDAW and an inquiry procedure in cases of systematic and widespread violations of the treaty. Although the Optional Protocol has 102 parties, these mechanisms have been invoked much less frequently than comparable provisions in other human rights treaties. CEDAW's website records that it has made a single inquiry under article 8 of the Optional Protocol in July 2004 dealing with violence against women in Mexico. As of 1 October 2010, the committee had received 27 communications alleging a breach of the Convention; by 1 October 2011, 10 of these had been found admissible and eight inadmissible.[12] Through its Concluding Observations on states' reports under the treaty, however, the CEDAW Committee has played a significant role in elaborating the Women's Convention. It has been very active in developing a jurisprudence of sex discrimination, emphasizing direct and indirect discrimination against women, and focusing on the reality of women's lives and the need for structural change, rather than simply formal equality.[13]

Overall the development of international human rights standards dealing with women illustrates a strategic dilemma, identified by Olympe de Gouges in the eighteenth century as the paradox of feminism: whether women's rights are best protected through general norms or through specific norms applicable only to women.[14] This dilemma pervades modern international legal responses to the unequal position of women: the attempt to improve women's lives through general laws can allow women's concerns to be submerged in what are deemed more global issues; however, the price of creating separate institutional mechanisms for women has been the building of a women's ghetto with less power, resources, and priority than the "general" human rights bodies.[15]

Feminist critiques have had some effect on the international human rights system.[16] One example, discussed in other chapters in this book, is the development of "gender mainstreaming" as a technique to encourage consideration of women's lives by the UN human rights machinery.[17] This concept was defined by the UN's Economic and Social Council (ECOSOC) in 1997 in the following way:

> Mainstreaming a gender perspective is the process of assessing the implications for women and men of any planned action, including legislation, policies or programmes, in all areas and at all levels. It is a strategy for making women's as well as men's concerns and experiences an integral dimension of the design, implementation, monitoring and evaluation of policies and programmes in all political, economic and societal spheres so that women and men benefit equally and inequality is not perpetuated. The ultimate goal is to achieve gender equality.[18]

Approaches to mainstream gender perspectives in the UN human rights system were developed in 1995 by a meeting of experts.[19] The Commission on Human Rights then adopted several resolutions over a number of years on "Integrating the Human Rights of Women Throughout the United Nations System."[20] These resolutions called for all components of the UN human rights system to "regularly and systematically take a gender perspective into account in the implementation of their mandates."[21] The resolutions specified the responsibility of the human rights treaty bodies to integrate a gender perspective, with reference to "gender-sensitive guidelines" in reviewing states' reports, preparing general comments and issuing recommendations and concluding observations. ECOSOC and the Commission on Human Rights also requested that the country-specific and thematic special rapporteurs, experts, and working groups include sex-disaggregated data in their reports, to address women-specific violations of human rights and to cooperate and exchange information with the Special Rapporteur on Violence against Women.[22] This is also an annual agenda item in the commission's replacement body, the Human Rights Council.[23]

The response to calls for gender mainstreaming in the UN human rights system has been rather muted, however.[24] This is to some extent the product of the low representation of women in most parts of the system. Responsiveness to the gender mainstreaming mandate seems to depend on the presence of at least one or two committee members who have a strong commitment to the issue. For example, the Committee on Economic, Social, and Cultural Rights has addressed the task of gender mainstreaming to a limited degree, at least insofar as it refers to the position of women in its Concluding Observations on States parties' reports and in its General Comments.[25] Gender mainstreaming appears in these contexts as exhortations to non-discrimination on the basis of sex and the inclusion of women in relevant decision-making, and as references to the special burdens women face with respect to access to the right in question. The Committee's reporting guidelines are uneven, however, with respect to women and gender issues. Some guidelines request information on the situation of women, but others do not. For example, the guidelines do not refer to the position of girls with respect to the right to free primary education.[26]

The Human Rights Committee, which monitors the ICCPR, has adopted some General Comments on articles of the treaty that are relevant to the position of women. The most explicit is General Comment 28 on Article 3 of the ICCPR, adopted in 2000.[27] This document examines each of the rights set out in the Covenant and analyzes the way in which they might affect women's lives. The interpretation and implementation of these provisions has, however, been circumscribed with respect to women. For example, the Human Rights Committee was slow to recognize indirect discrimination against women, a significant impediment to women's advancement.[28] In the context of the international prohibition of torture, Alice Edwards has shown how the UN treaty bodies have a mixed record in applying general principles to women's lives.[29]

The Human Rights Committee has occasionally used its adoption of Concluding Observations to raise concerns about women's lives. For example, Concluding Comments on Peru's 1996 periodic report under the ICCPR drew attention to the criminalization of abortion even in cases of rape, which had resulted in illegal abortions becoming the major cause of maternal mortality. The committee stated that "these provisions not only mean that women are subject to inhumane treatment but are possibly incompatible with articles 3 [the right of men and women to equal enjoyment of the rights set out in the ICCPR], 6 [the right to life] and 7 [the right to be free from torture and cruel, degrading and inhuman treatment] of the Covenant."[30] Generally, however, the Human Rights Committee confines itself to occasional questions and comments about the number of women in public life and seeks more statistical information on women.[31] There is little evidence of more probing questions about the impact of gender on the enjoyment of human rights.

The records of the annual meetings of the chairpersons of the human rights treaty bodies suggest that there is no regular discussion of gender mainstreaming.[32] Guidelines on States parties' reports to the treaty bodies refer only briefly to the strategy.[33] The reports of the special procedures of the Human Rights Council suggest a resistance to, or misunderstanding of, gender mainstreaming. At best, they may give information about individual cases where women were victims of human rights abuses, but there is no analysis of the relationship between these harms and women's status in public and private life.[34] The special rapporteurs also tend to refer to the category of "women and children," reinforcing women's identity and value as mothers as well as positioning them as inherently vulnerable. Gender mainstreaming is most often understood as requiring attention to how many women were affected by the right or situation in question. Information about women, for example, in the reports of successive special rapporteurs on the Situation of Human Rights in Afghanistan, was brief and broad-brushed, involving general statements without any analysis.[35] This pattern of overlooking women may be linked to the fact that there are many more men than women appointed as "mainstream" special rapporteurs.[36] By far the most significant special rapporteur with respect to women and gender is the specialized Special Rapporteur on the Elimination of Violence against Women, and some of the Human Rights Council's other special procedures tend to shift responsibility for considering women's lives onto this one office. The Human Rights Council has, however, recently established a Working Group on Discrimination against Women in Law and Practice which may deepen the Council's approach.[37]

Overall, attention to questions of sex and gender in the UN human rights system has been haphazard. At best, there is attention to the position of women in particular contexts, mainly in statistical terms, but there has been little understanding of the way in which stereotypes about sex and gender roles can affect the human right in question. Violations of women's human rights are typically presented as an aspect of women's inherent vulnerability, as if this attribute were a biological fact.

Gender as a concept emerged somewhat later than sex in international legal instruments, but has rapidly become the preferred term, swallowing up the idea of sex. This elision is well illustrated by the definition of gender used in the Statute of the International Criminal Court:

> For the purpose of this Statute, it is understood that the term "gender" refers to the two sexes, male and female, within the context of society. The term "gender" does not indicate any meaning different from the above.[38]

The elision causes a number of problems. First, it links gender with biology, implying that gender is a fixed, objective fact about a person. It does not capture the ways that gender is constructed in society to make some actions seem natural and others controversial. It reaffirms the "naturalness" of female/male identities and bypasses the performative aspects of gender. Understanding gender as essentially about biological sex does not capture the relational nature of gender, the role of power relations and the way that structures of subordination are reproduced.[39] It allows problems facing women to be understood as the product of particular cultures, lack of participation in public arenas or lack of information or skills and obscures the way that gender shapes our understanding of the world. Most significantly, the term "gender" has become a synonym for women in the international arena, leaving both the roles of men and male gender identities unexamined, as though they were somehow natural and immutable.

The concept of gender in international human rights law is animated by an idea of equality as equal treatment of women and men, assuming symmetry of position between women and men. It does not address the complex ways that gender is created and sustained by social and power relations.[40] Treating women and men as though they face similar obstacles perpetuates existing disparities between them. In some accounts of gender mainstreaming, the strategy has become simply a head count of women in particular positions, a modest variation of the "equal opportunity" agenda.[41] While increasing women's participation in institutions is important, it does not itself change institutional agendas.

Moreover, the definition of gender mainstreaming in international institutions contemplates a limited sphere for its operation. It is regarded as primarily relevant to policy development in particular areas, such as development, human rights and some aspects of labor markets. Other fields appear immune to gendered scrutiny, for example competition policy.[42] Within the UN, most areas of law have been treated as if they were impervious to concerns of gender: gender mainstreaming mandates have not been given to either the International Law Commission or the International Court of Justice.

My point here is that the accounts of sex and gender deployed in international law are restricted and the two terms are largely taken as interchangeable. They can be summarized as prohibiting direct, or formal, discrimination against women, and, on occasion, some forms of indirect discrimination. They are taken to apply primarily in the public spheres of politics, education and the economy and to be irrelevant or insignificant in other areas.

Women's rights after conflict

The role of international law in post-conflict situations has prompted considerable attention over the last 20 years. State-building relies on the human rights concept of self-determination. In international law, self-determination accords a people the right to autonomy, freedom from alien oppression and the right to choose an economic, political and social system "free from outside intervention, subversion, coercion or constraint of any kind whatsoever."[43] Once external self-determination has been achieved and internal self-determination is guaranteed, it is assumed that all members of the group will equally benefit, in other words that the terms "self" and "peoples" are homogenous. Individual and group aspirations and goals, both before and after the achievement of self-determination, are subsumed within those of the self-determining unit. This assumption of group identity and commonality disadvantages many women. The notion of a self-determining unit collapses many forms of diversity, but most particularly that of sex. The consequences of this limited definition are evident in the fact that apparently successful claims to self-determination typically fail to deliver the same level of personal freedom and autonomy for women as for men. Indeed, in many cases achievement of national self-determination has led to deterioration in the position of women.

The UN Security Council's adoption of Resolution 1325 in 2000 drew attention to the significance of involving women in peacebuilding, but its promise has not been realized. The language of women's rights can be undermined on the one hand by crude applications of the concepts without attention to local contexts and on the other by a willingness to trade off women's rights to achieve an apparent political stability. In the lead up to the invasion of Afghanistan in 2001, for example, women's rights were regularly used by the West as a rationale for the conflict and state-building after conflict was presented as an opportunity to improve women's lives. Thus the UN Development Fund for Women (UNIFEM) (now part of UN Women) defined its role in the project of reconstruction after the conflict in Afghanistan as "advanc[ing] gender equality by supporting new legislation, backing women's leadership and equal representation, and widening the space for women's participation at peace tables."[44] Sari Kouvo's study of women in Afghanistan suggests that these goals have been approached in an unsatisfactory manner, without taking into account the specificity of the situations they are addressing; for example, assuming that women in Afghanistan form a homogenous group, that particular women can represent all Afghan women, and that women will be able to influence political decisions. Indeed, she argues, emphasis on women's participation and gender mainstreaming has created a backlash in Afghanistan: those hostile to the policies supporting women have argued successfully that they are the product of foreign influences seeking to undermine Afghan culture and Islam.[45]

In other contexts, local campaigns for equality have met with hostility. In East Timor, emerging from the long Indonesian occupation into independence, gender roles assigning men to a public world of politics and employment and women to a private world of home and family pervade social and economic relations. They are supported by religious doctrine, low levels of education and traditional practices.[46] The invocation of the vocabulary of women's human rights has caused tension in this context. Thus the East Timorese resistance (and now political) leader, Xanana Gusmão, complained in 2001 of the "obsessive acculturation to standards that hundreds of international experts try to convey to the East Timorese, who are hungry for values."[47] He implied that the right of women to determine their own lives had no natural affinity with East Timorese culture. This type of argument has been strongly resisted by women's groups in East Timor, but has continued to undermine women's rights campaigns.[48]

Iraq provides another example. Despite the oppressive nature of Saddam Hussein's regime, Iraqi women generally benefitted from its secular approach to personal status laws.[49] The UN Arab Development Report noted in 2002 that Iraqi women rated highest among Arab women on UN measures of "gender empowerment" because of their significant political participation.[50] The 2003 invasion of Iraq, among other things, has allowed an assertive masculine traditionalism to re-claim authenticity and authority. There has been considerable local resistance, particularly from religious groups, to international pressure for the involvement of women in public life.[51] The international community has often tolerated these culturally based rejections of women's rights to prevent alienating local leaders.

Another concern for women in state-building is the way that relevant international human rights standards are defined. Representation of women has been the major concern addressed, albeit superficially and inadequately, by the international community, with intense debate about the legitimacy and impact of quotas for women. Other problems women face tend to be obscured. For example, Sumie Nakaya points to the lack of attention paid to the effect of the ethnic partitions model used in state-building in Bosnia-Herzegovina and Kosovo where women who had married across ethnic lines faced both sex and race discrimination.[52] Moreover the right to equality in international law invoked in state-building has essentially remained tethered to a limited, procedural, account of non-discrimination which has had minimal effect on improving women's lives.

A related problem is the tendency in state-building to emphasize civil and political rights over economic, social and cultural rights.[53] This is of course an issue for men as well as women, but existing discrimination against women in areas such as access to land or inheritance of housing and property makes it particularly acute for women.[54] If the empowerment of women is understood to depend simply on the inclusion of women in various spheres of public life, there will be inadequate attention to the gendered nature of the rules of the game that women are required to play. International institutions need to rethink traditional structures of public and private life, such as ideas of economic activity and value, to accommodate women's lives. We also must consider the effects of the unequal balance of domestic labor which limits the capacity of women to operate in the public, political sphere.

While the vocabulary of women's rights is now a staple of international peacebuilding packages, women's rights are often traded away in the search for stability and strong leadership. There is often a "selective mapping" of feminist concepts by international actors, as well as a failure to engage with their intended beneficiaries.[55]

Conclusion

Is international human rights law a useful mechanism to disseminate feminist norms? In some contexts, it seems that the basic principle of equality between women and men has not traveled at all beyond the international instruments that enshrine it. Even when international law relating to sex and gender has become part of national legal systems, it is often under challenge from local assertions of male culture and international diffidence.

The limited efficacy of international human rights law in this context suggests that UN Women could be a valuable focal point for research, training and initiatives to improve women's lives, although we need also to recognize the risk that the creation of such an entity may allow sex and gender to be sidelined from mainstream international concerns and effectively ignored.[56] An important task for UN Women will be to move the debate about women's lives from one focused on the norms of non-discrimination and equality with men. The language of non-discrimination and equality contained in international human rights instruments has achieved little for women, as it is easily reduced to ideas of formal equality. The promise of non-discrimination against women when compared to similarly situated men has a blunt edge because, across the world, women's lives differ in so many ways from men's. Reproduction, economic opportunities, education, and political and social contexts, for example, all mark women's lives as distinct to those of men.[57] The concept of de facto equality, which is endorsed by the Women's Convention, is designed to deliver a substantive form of equality. But this has proved as hard to achieve in the international sphere as it has in national legal systems; an equality discourse makes it difficult to emphasize the particularity of women's lives in ways that benefit women. It can mask the details of women's lives and force them into cumbersome and inaccurate categories. One example of this is the international legal understanding of violence against women as a form of discrimination. In practice, violence is less a product of discrimination than it is of power and domination, and understanding it as discrimination can lead to strained and unconvincing arguments.

One challenge for UN Women in the field of human rights is to develop concepts that capture the limits of women's lives and create political momentum for change.[58] Examples of such concepts in national legal systems include that of sexual harassment. As Rosemary Hunter has pointed out, the notion of sexual harassment "graphically specifies what is involved, connotes injustice, and presses for intervention both to end the practice and to provide compensation for those who have been harassed."[59] She proposes alternatives to discrimination paradigms, such as "undervaluation" in the context of pay disparities for women rather than that of "equal pay," which immediately implicates a male comparator. Another alternative to the language of discrimination is that of "policy neglect" in the area of resource allocation, which draws attention to groups that end up disadvantaged by apparently neutral policies. These ideas have resonance in the international arena; for example the concept of policy neglect could be used to capture the effect of international monetary institutions' policies on women's lives.[60] The Women's Convention includes some valuable but underused provisions that do not rest on a male comparator, for example the requirement that states take steps to "modify the social and cultural patterns of conduct of men and women, with a view to achieving the elimination of prejudices and customary and all other practices which are based on the idea of the inferiority or the superiority of either of the sexes or on stereotyped roles for men and women."[61]

Olympe de Gouges's conundrum about the best strategy to pursue women's rights cannot be easily resolved. Women are always likely to struggle for recognition of their rights whether they operate within general or women-specific agencies. The way ahead is to work through all available institutions to lessen the disparity between the lives of women and men across the globe. Breaking away from the limited equality paradigm endorsed by international law would be a good beginning.

Feminist commitments, such as the equality of women, have influenced the development of international human rights law, but they have been incorporated only in a partial manner and implemented without regard to context. This underlines a distinction between feminist messages and feminist methods in international law. The former have been influential in rhetorical terms, while the latter have been under-explored. Feminist messages, however, are likely to be productive only if they are deployed through feminist techniques such as "world traveling."[62] This involves being explicit about our own historical and cultural background, trying to understand how other women might see us, and recognizing the complexities of the lives of other women.[63] Feminist methodologies suggest that sex and gender norms must respond to the needs and desires of women. Understanding these needs is not always easy and requires patience and empathy, particularly in post-conflict situations.

Thirty-three years after the adoption of the Women's Convention and after more than 20 years of feminist scholarship in international law, we can observe that some sex and gender norms have arrived in international law, although they can be hollowed out by bland repetition and deployed to reinforce the status quo. But the language of justice and international law contains radical potential, even if it is realized only occasionally. International human rights law in particular, as Sally Engle Merry has observed, "is always in danger of escaping its bounds and working in a genuinely emancipatory way."[64] The most valuable aspect of international human rights law as a portmanteau for feminist norms may thus be through its capacity to articulate particular political goals and engender cultural change.

Notes

1 *Charter of the United Nations*, 26 June 1945, Article 1.
2 Ibid., Article 8.
3 ICCPR, Article 2; ICESCR, Article 2.
4 ICCPR, Article 3; ICESCR, Article 3.
5 ICCPR, Article 26.
6 Ibid., Article 2.
7 Ibid., Articles 7, 8, 10, 11, 12, 13b, 13c, 14, and 15.
8 Ibid., Article 16.
9 General Recommendation 19, *Violence against Women* (CEDAW document A/47/38, CEDAW/C/1992/L.1/Add 15), 24 June 1992.
10 The text of reservations and objections to the reservations is available at: United Nations Treaty Collection 2012, *Chapter IV, Human Rights, Convention on the Elimination of All Forms of Discrimination Against Women*, New York, 18 December 1979, http://treaties.un.org/Pages/ViewDetails.aspx?src=TREATY&mtdsg_no=IV-8&chapter=4&lang=en.
11 Vienna Convention on the Law of Treaties, Article 19. But note that the UN International Law Commission has proposed a new scheme for dealing with reservations, to be adopted in 2011: see draft guidelines of the Guide to Practice on Reservations to Treaties at International Law Commission 2011, *Reservations to Treaties*, http://untreaty.un.org/ilc/summaries/1_8.htm.
12 See OHCHR 2012, *Optional Protocol to the Convention on the Elimination of Discrimination against Women*, http://www2.ohchr.org/english/law/cedaw-one.htm.
13 See, e.g., General Recommendation 25, *Temporary Special Measures* (General Assembly document A/59/38), 12–30 January and 6–23 July 2004, para 4. Paragraph 10 calls on states parties to reformulate laws, policies and practices "so that they are no longer grounded in historically determined male paradigms of power and life patterns."
14 Sari Kouvo, *Making Just Rights? Mainstreaming Women's Human Rights and a Gender Perspective* (Uppsala: Iustus förlag 2004), 104.
15 See Hilary Charlesworth and Christine Chinkin, *The Boundaries of International Law: A Feminist Analysis* (Manchester: Manchester University Press, 2000), 219.
16 See generally Rachel Johnstone, "Feminist Influences on the United Nations Human Rights Treaty Bodies," *Human Rights Quarterly* 28 (2006): 148–85.
17 Hilary Charlesworth, "Not Waving but Drowning: Gender Mainstreaming and Human Rights,"*Harvard Human Rights Journal* 18 (2005): 1.
18 United Nations, *Platform for Action, ECOSOC Agreed Conclusions* 1997/2 (1997), http://www.un.org/womenwatch/osagi/pdf/ECOSOCAC1997.2.PDF.
19 *Expert Group Meeting on the Development of Guidelines for the Integration of Gender Perspectives into Human Rights Activities and Programmes* (ECOSOC document E/CN.4/1996/105), 20 November 1995.

20 *Integrating the Human Rights of Women throughout the United Nations System* (OHCHR document E/CN.4/RES/1998/51), 17 April 1998; *Integrating the Human Rights of Women throughout the United Nations System* (OHCHR document E/CN.4/RES/1999/41), 24 April 1999; *Human Rights Defenders* (OHCHR document E/CN.4/RES/2000/61), 27 April 2000.

21 *Integrating the Human Rights of Women throughout the United Nations System* (OHCHR document E/CN.4/RES/1997/43), 15 April 1997.

22 For details of progress in this regard, see *The Question of Integrating the Human Rights of Women throughout the United Nations System: Report of the Secretary-General* (OHCHR document E/CN.4/2005/68), 10 January 2005, at paragraphs 43–53.

23 See Human Rights Council Resolution 5/1 (18 June 2007) on the principle of the integration of gender perspectives in the Council's work.

24 For details of the response of all the human rights treaty bodies up to 2005, see *The Question of Integrating the Human Rights of Women throughout the United Nations System: Report of the Secretary-General* (OHCHR document E/CN.4/2005/68), at paragraphs 23–41. For a study of the period 1995–1997, see Anne Gallagher, "Ending the Marginalization: Strategies for Incorporating Women into the United Nations Human Rights System," *Human Rights Quarterly* 19 (1997): 283. For a survey of six of the treaty bodies' work in this area up until 2003, see Kouvo, *Making Just Rights?*, 257–82. See generally Alice Edwards, *Violence Against Women under International Human Rights Law* (Cambridge: Cambridge University Press 2011), 88–139.

25 See, e.g., *General Comment No. 20 on Non-discrimination in Economic, Social and Cultural Rights* (ECOSOC document E/C.12/GC/20), 2 July 2009. See Kouvo, *Making Just Rights?*, 268–71 for analysis of the work of the committee.

26 See generally Gallagher, "Ending the Marginalization," 301–2.

27 *General Comment 28 on Article 3 of the ICCPR: Equality of Rights Between Men and Women* (CCPR document CCPR/C/21/Rev.1/Add.10), 29 March 2000. See also *General Comment 20 on Article 2, Paragraph 2 of the ICESCR: Non-discrimination in Economic, Social and Cultural Rights* (ECOSOC document E/C.12/GC/20).

28 For example, *Cavalcanti Araujo-Jongen v The Netherlands* 418/1990; *A.P.L. v d.M v The Netherlands* 478/1991. Compare *Broeks v The Netherlands* 172/1984.

29 Alice Edwards, "The 'Feminizing' of Torture under International Human Rights Law," *Leiden Journal of International Law* 19 (2006): 350.

30 Christine Ainetter Brautigam, "Mainstreaming a Gender Perspective in the Work of the United Nations Human Rights Treaty Bodies," *American Society of International Law Proceedings* 91 (1997): 389, 393.

31 See, e.g., *International Covenant on Civil and Political Rights, List of Issues to Be Taken Up in Connection with the Consideration of the Fifth Periodic Report of Morocco* (CCPR document CCPR/C/82/L/MAR), 13 August 2004.

32 See, e.g., *Report of the Chairpersons of the Human Rights Treaty Bodies on their Twentieth Meeting, 63rd Session, Provisional Agenda Item 67(a)* (General Assembly document A/63/280), 13 August 2008. There is a brief reference to statistics on gender at paragraph 32.

33 *Compilation of Guidelines on the Form and Content of Reports to be Submitted by States Parties to the International Human Rights Treaties, Report of the*

Secretary-General (HRI document HRI/GEN/2/Rev.5), 29 May 2008, paragraph 42 (f).

34 See Human Rights Council Resolution 6/30, 14 December 2007, on integrating the human rights of women throughout the United Nations system, and Human Rights Council Resolution 12/17, 2 October 2010, on discrimination against women, suggesting that progress is slow.

35 Kouvo, *Making Just Rights?*, 235–36, 248–51.

36 Of 36 special rapporteurs and independent experts listed on the website of the Human Rights Council as at 1 July 2011, 11 are women and 25 are men.

37 Human Rights Council Resolution 15/23, 8 October 2010, on the elimination of discrimination against women.

38 Article 7.3.

39 Susanne Baden and Anne-Marie Goetz, "Who Needs [Sex] When You Can Have [Gender]?" *Feminist Review* 56 (1997): 3, 7.

40 Elizabeth Reid, "Transformational Development and the Wellbeing of Women," *Development Bulletin* 64 (2004): 19.

41 Alison Woodward, "Gender Mainstreaming in European Policy: Innovation or Deception?" Discussion Paper FS 101–3 Wissensschaftszentrum Berlin fur Sozialforschung, October 2001, 22.

42 Mark A. Pollack and Emilie Marie Hafner-Burton, "Mainstreaming Gender in the European Union," *Journal of European Public Policy* 7 (2000): 432, 446.

43 The wording is taken from a General Assembly Resolution on Afghanistan after the Soviet invasion in 1979. General Assembly Resolution ES-6 2, 14 January 1980.

44 Available at www.unifem.org/gender_issues/governance_peace_security/at_a_glance.php.

45 Sari Kouvo, "Taking Women Seriously? Conflict, State-building and Gender in Afghanistan," in *Feminist Perspectives on Contemporary International Law: Between Resistance and Compliance?*, ed. Sari Kouvo and Zoe Pearson (Oxford: Hart Publishing, 2011), 159–76.

46 See Susan Harris Rimmer, *Gender and Transitional Justice: The Women of East Timor* (New York: Routledge, 2010).

47 Kay Rala Xanana Gusmão, 31 December 2000, *New Year's Message*, www.pcug.org.au/~wildwood/JanNewYear.htm.

48 See Nina Hall and Jacqui True, "Gender Mainstreaming in a Post-conflict State: Towards Democratic Peace in Timor-Leste?" in *Gender and Global Politics in the Asia*-Pacific, ed. Bina D'Costa and Katrina Lee-Koo (New York: Palgrave Macmillan, 2009), 159.

49 Gina Tabet, "Women in Personal Status Laws: Iraq, Jordan, Lebanon, Palestine, Syria," SHS Papers in Women's Studies/Gender Research, 4 (Paris: UNESCO, July 2005), 11–12.

50 United Nations Development Programme, *Arab Human Development Report: Creating Opportunities for Future Generations* (New York: United Nations Publications, 2002). In 2002, women held almost 20 percent of parliamentary seats in Iraq compared to the 3.5 percent average for Arab states.

51 Swanee Hunt and Cristina Posa, "Iraq's Excluded Women," *Foreign Policy* (July/August 2004).

52 Sumie Nakaya, "Women and Gender Equality in Peacebuilding: Somalia and Mozambique," in *Building Sustainable Peace*, ed.Tom Keating and W. Andy Knight (Tokyo: UN University Press, 2004), 143, 146.

53 Ismael Muvingi, "Sitting on Powder Kegs: Socioeconomic Rights in Transitional Societies," *International Journal of Transitional Justice* 3 (2009): 163; Hilary Charlesworth, "Human Rights and the Rule of Law After Conflict," in *The Hart-Fuller Debate in the Twenty-First Century*, ed. Peter Cane (Oxford: Hart Publishing, 2010), 43.

54 See Nora Niland, "Rights, Rhetoric and Reality: A Snapshot from Afghanistan," in *The UN, Human Rights and Post-Conflict Situations*, ed. Nigel D. White and Dirk Klaasen (Manchester: Manchester University Press, 2005), 322; Christine Chinkin and Hilary Charlesworth, "Building Women into Peace: The International Legal Framework," *Third World Quarterly* 27 (2006): 937.

55 See Dianne Otto, "Remapping Crisis Through a Feminist Lens," in *Feminist Perspectives on Contemporary International Law: Between Resistance and Compliance?* ed. Sari Kouvo and Zoe Pearson (Oxford: Hart Publishing, 2011), 75–95, 86–88.

56 This type of argument was put by Rachel Mayanja, Special Adviser to the UN Secretary-General on Gender Issues. She argued that a women's agency could "create a ghetto for women's issues," and that instead women's issues should be given equal weight within existing UN agencies. See *Toronto Star*, 1 July 2006.

57 See Rosemary Hunter, "Introduction: Feminism and Equality," in *Rethinking Equality Projects in Law: Feminist Challenges*, ed. Rosemary Hunter (Oxford: Hart Publishing, 2008), 4.

58 Rosemary Hunter makes this proposal in the context of national legal systems, ibid., 8.

59 Ibid., 82–83.

60 See the work of Isabella Bakker on "strategic silences" about women in macroeconomic policies: *The Strategic Silence: Gender and Economic Policy* (London: Zed Press/The North-South Institute, 1994); Brigitte Young, Isabella Bakker and Diane Elson, ed., *Questioning Financial Governance from a Feminist Perspective* (London: Routledge, 2011).

61 Article 5a. The possibilities of this provision are explored in Rebecca J. Cook and Simone Cusack, *Gender Stereotyping: Transnational Legal Perspectives* (Philadelphia, PA: University of Pennsylvania Press, 2010).

62 The term was introduced by Maria Lugones in "Playfulness, 'World-traveling', and Loving Perception," *Hypatia* 2 (1987): 3.

63 Isabelle Gunning, "Arrogant Perception, World-Traveling and Multicultural Feminism: The Case of Female Genital Surgeries," *Columbia Human Rights Law Review* 23 (1991–92): 189, 191. See also Ann McNevin, "Confessions of a Failed Feminist IR Scholar: Feminist Methodologies in Practice in Peshawar," in *Gender and Global Politics in the Asia-Pacific*, ed. Bina D'Costa and Katrina Lee-Koo (New York: Palgrave Macmillan, 2009), 115.

64 Sally Engle Merry, *Human Rights and Gender Violence: Translating International Law into Local Justice* (Chicago: Chicago University Press, 2006), 231.

2 Gender mainstreaming strategies in international governance

Jacqui True and Laura Parisi

- **Gender mainstreaming: A review**
- **Critically evaluating gender mainstreaming**
- **Conclusion**

A significant "gender mainstreaming" policy revolution was sparked by the 1995 Beijing Platform for Action (BPA), ratified by member states at the United Nations Fourth World Conference on Women. The platform clearly articulated the need for "gender mainstreaming" across all policy areas at all levels of governance. It marked a critical turning point in global governance and development of the gender equality regime, although the 1993 United Nations Vienna Declaration on Human Rights is widely touted as the first "gender-mainstreamed" human rights document for its recognition of women's rights as human rights.[1] The BPA called on governments and other governance actors to incorporate a gender analysis into all programs and policies in order to assess the differential gender impacts of those policies on women and men.[2]

Implicit in this gender mainstreaming directive is the idea that women and men, due to socially-constructed gender differences and ideologies, do not experience political, economic, social, and cultural life in the same ways. United Nations (UN) member states recognized that if policies fail to take these differences into account, they are unlikely to be effective or to meet their intended goals, and gender injustices and inequalities affecting both women and men will be perpetuated. For example, the right to be free from torture conjures up images of men as political prisoners being tortured by government actors in the public sphere. It has historically failed to address the forms of torture women may experience in the private sphere due to domestic abuse and violence making them prisoners in their own homes. Gender mainstreaming practices and policies allow us to consider all the forms of torture that an anti-violence initiative needs to take into account. Moreover, for feminists a central goal of gender mainstreaming is to transform existing gender relations, to reduce (and hopefully eliminate) gender inequality that creates injustices for women and men, and to foster women's empowerment given their lesser access to key social and economic resources and political rights.[3]

In 1997, the BPA's definition of gender mainstreaming became the basis for official system-wide policy development in the UN, conceptualized in the 1997/2 Economic and Social Council (ECOSOC) Agreed Conclusions.[4] The World Bank officially adopted gender mainstreaming as a policy directive in 2001, though it had been making some efforts to mainstream gender issues since the mid-1990s, while the International Labour Organization (ILO) officially prioritized gender mainstreaming in the late 1990s, though it has a long history of recognizing women's labor rights, most recently the rights of domestic workers.[5] At the same time, there was also a global diffusion of gender mainstreaming as a policy norm at the state/domestic level as evidenced by a significant increase in women's policy machineries with mainstreaming mandates.[6] This institutional/norm diffusion also spread to regional governance structures, such as the European Union (EU) and Asia Pacific Economic Cooperation (APEC), as well to nongovernmental organizations (NGOs).[7]

More than 15 years after its official adoption at the UN, what is the state of gender mainstreaming policies and practices in global governance? In this chapter, we review the conceptual basis of gender mainstreaming, identify the forms of gender mainstreaming that are currently operating, and evaluate some of the common pitfalls of gender mainstreaming in international organizations. We conclude by assessing whether or not gender mainstreaming is an effective feminist governance strategy and what it might mean to move beyond current, often ineffective, practices of gender mainstreaming.

Gender mainstreaming: a review

Existing scholarship on gender mainstreaming tends to fall into three main areas, which are overlapping and not mutually exclusive. The first area is concerned with theoretical and definitional debates regarding the meanings of gender, gender equality, and gender mainstreaming, which have implications for organizational practices and programming.[8] Some of these studies approach mainstreaming from a discursive perspective that is concerned with the forms of power it produces through policy processes and the diffusion of an international strategic language that creates new subjects and objects. We expand on this discussion later in the chapter.

The second area of gender mainstreaming literature generally consists of in-depth case studies of particular NGOs,[9] national women's policy machineries and governmental practices,[10] regional governance structures,[11] and intergovernmental organizations (IGOs).[12] Many of these studies make distinctions between *institutional* and *operational* gender mainstreaming strategies, which emphasize the need to focus on the internal dynamics of the organization as well as the need to change programming and policy initiatives, respectively.[13] These studies also focus on the varied translation of global mainstreaming norms in local, national or regional policymaking settings and the degree of engagement between women's or feminist movements and gender mainstreaming organizations.

The third area focuses on what political factors or economic incentives need to be in place for organizations/governments to pursue gender mainstreaming strategies.[14] A related literature analyzes those organizations which have adopted gender mainstreaming initiatives and how to evaluate the success or failure of these initiatives.[15] In reviewing the literature what becomes evident is that governance organizations are interpreting and implementing gender mainstreaming mandates in diverse ways. As a result, a variety of gender mainstreaming models or approaches have emerged transnationally. These approaches are not necessarily mutually exclusive and can be present simultaneously with an organization or governance structure.[16]

1. *Gender equality or integrationist model*: Organizations/governments may be adopting integrationist models, which add gender concerns to existing frameworks without fundamentally altering those frameworks. This is often termed the "add women and stir approach."[17] In this case, the concept of gender becomes synonymous with women and the basic purposes and operating procedures of institutions and organizations remain unchanged.[18]

2. *Gender-as-difference model*: This model is related to the integrationist model in that it keeps the gender binary firmly entrenched without displacing masculinist power structures.[19] The focus on and reification of gender difference invariably relies on a notion of tolerance for difference, in this case women and/or femininity in relation to masculinist norms, and this difference is what is being mainstreamed.[20] This model is a form of integration though it can take the form of women-only programming, recognizing the value of marginalized voices and group perspectives.[21]

3. *Gender-as-intersectionality model*: Organizations may be contextualizing and adapting gender mainstreaming strategies to suit their particular needs, taking into account the multiple and intersecting identities of the populations that they serve.[22]

4. *Alternative gender mainstreaming models or transformationalist model*: Another possibility is that organizations are generating entirely new conceptions of gender mainstreaming. For example, Brooke Ackerly argues that women's human rights activists (and by extension NGOs) are capable of generating new understandings of gender equality and women's human rights.[23] Transformationalist models call for using a gender perspective to completely reform or reorient existing global governance frameworks, for instance with respect to human rights and development, with relevance across organizations/governments.[24]

5. *Resistance to/Rejection of gender mainstreaming model*: Organizations and institutions may also be resistant to implementing gender mainstreaming because it is too confusing or too threatening, resulting in superficial policies that are in name only or represent the rejection of gender mainstreaming mandates altogether.[25]

Critically evaluating gender mainstreaming

Gender mainstreaming was conceived by feminists as an international strategy for bringing about greater gender equality through policy-making. But in practice it has often served to diminish and depoliticize gender equality goals. As the various organizational models in circulation indicate, mainstreaming is a practice of power itself as much as it seeks to transform gendered relations of power. In this part of the chapter we evaluate gender mainstreaming, reflecting on five common criticisms of gender mainstreaming made by feminist scholars supported by research in international development and security areas. These criticisms argue 1) that resistance to gender mainstreaming by institutional actors, as in model 5 above, undermines its intended effects; 2) that gender mainstreaming is often based on a single normative perspective on gender, as a synonym for women thus reinforcing gender stereotypes as is the case in models 1 and 2 above; 3) that the gender equality impacts and outcomes of mainstreaming are at best challenging to monitor and evaluate; 4) that women's movements and civil society have been largely excluded from the development of mainstreaming policy; and finally, 5) that the gender mainstreaming agenda has been variously co-opted as a means to other institutional ends such as those of state security and economic growth. Taken together, we argue that there are demonstrable weaknesses in both the concept and the implementation of gender mainstreaming. These weaknesses will need to be addressed by organizations and activists if mainstreaming is effectively to promote women's rights and global gender equality.

Resistance to mainstreaming

Gender mainstreaming has provoked both ambivalence and contradictory practice in international organizations. The failure to embrace gender mainstreaming in the operational work of development and security organizations has to do with a deep-seated misconception about what gender means. As model 5 above suggests, among policymakers there is a lack of understanding of the substantial significance of gender perspectives to the policy issues and work at hand despite all the technical tools and gender training developed to assist analysts. For instance, many government aid agencies complete the Organisation for Economic Co-operation and Development (OECD) Development Assistance Co-operation (DAC) gender policy marker "gender mainstreaming" checklist, which requires them to consider whether a development project has a gender dimension or impact, based on little knowledge, research or consultation with gender experts about the potential and often complex ways that women and men are affected differentially by development processes (as well as how efforts to increase gender equality may have important impacts on other development goals).[26] Yet, this self-reported "gender review" of overseas development aid (ODA) is aggregated to score and rank the gender mainstreaming performance of donor countries.[27]

Practice within organizations that contradicts gender equality principles also has to do with bureaucratic resistance to the adoption of yet another approach that must be mainstreamed or checked off without any overarching methodology for integrating it as implied by the gender-as-intersectionality model (3) above. Increasingly a range of cross-cutting issues, not just gender equality, must be addressed in policymaking processes including children's and youth rights, environmental impacts, diversity matters, cross-cultural sensitivity and so on. Moreover, gender mainstreaming may be conceived and deployed politically as a way to undermine transformative gender equality goals by eliminating specific budgets and personnel devoted to women or gender issues. Stratigaki documents the way gender mainstreaming led to the dissipation of any focus on gender inequalities in the Employment and Social Affairs Directorate-General of the European Commission, as do Debusscher and True with respect to the EU's Development Directorate-General.[28] In the latter case the requirement that gender analysis should be mainstreamed within all EU aid budgets increasingly under country ownership following the OECD Paris Declaration[29] eliminated the allocation of a proportion of total development aid for specific projects designed to increase gender equality and women's empowerment.

Many NGOs and developing states resist the gender mainstreaming mandates imposed on them by Northern funders, although they may be intended to address forms of violence against women or women's participation in peacebuilding.[30] For example, many UN member states have yet to adopt national action plans for the implementation of Security Council Resolution 1325 (see von Braunmühl, this volume), and the prospect of collective action remains bleak since no conflict-affected country will likely produce a national action plan unless other conflict countries do so as well.[31] Some conflict-affected countries actually perceive these national plans as tools of rich donor countries with the potential to apply conditionality to their development assistance and private investment funds. Indeed, the majority of the 37 states that have already created national action plans for implementing 1325 in their foreign and security policies are Northern, developed nations.[32] Thus the power relations among states and international agencies, bureaucratic delays, and deep organizational pathologies resistant to change all work to undermine gender mainstreaming efforts. Regardless of their transformative intentions, gender mainstreaming strategies frequently do not take account of these dynamics nor do they anticipate unintended consequences.

What is the gender in gender mainstreaming?

Gender mainstreaming has been frequently criticized for reflecting a single gender perspective, often based on Western, heterosexist norms of appropriate gender relations, and reflected in mainstreaming models 1 and 2 above. Kate Bedford's (also this volume) research on World Bank anti-poverty policy in Ecuador exemplifies this second line of feminist critique. She argues that the World Bank's promotion of a gender equal model of dual earner/dual care family households in anti-poverty policies is based on a norm of heterosexual marriage even while it claims to promote women's economic independence.[33] Gender is generally not mainstreamed in policy in an intersectional way along the lines of mainstreaming model 3 but is used as a synonym for the inclusion of women addressing the experiences of dominant rather than marginalized groups of women. This is partly due to the different forms of feminism that permeate the 1995 BPA, the key instrument in the international diffusion of gender mainstreaming. Though paragraph 46 of the platform makes a weak attempt at acknowledging diversity and attendant multiple discriminations, this "gender-as-intersectionality" discourse is accompanied by a strong liberal gender equality discourse in the document.[34]

Some current national approaches to gender mainstreaming understand the subjects of policy as diverse with respect to gender, ethnicity, race, sexuality, class, etc., and incorporate this diversity into the policymaking process by addressing gender, race, nationality status, and sexuality differences together rather than separately.[35] However, such an approach has hardly made any traction within international organizations still grappling with the differences between women and men, and, in the process, falling into the trap of stereotyping these differences as if they were biologically fixed rather than socially and politically defined. For instance, the UN Security Council in its Resolution 1325 (2000) recognizes the need for integrating a gender perspective in planning peace and security operations. However, it has focused primarily on the greater participation in peacekeeping roles of women[36] who are expected to "civilize" missions.[37] The resolution does not broach the causes of conflict in the culture of masculine aggression promoted in militaries whether engaged in armed conflict or peacekeeping missions or discuss the disproportionate impact of peacekeeping operations on women due to gendered socio-economic inequalities that make them more vulnerable than men civilians after conflict.[38] These impacts include the many more women than men internally displaced after conflict, the ongoing post-conflict issues of violence against women, including abuse by local and international peacekeepers, and sex trafficking.[39]

Moreover, the implementation of mainstreaming in the UN Department of Peacekeeping Operations (DPKO) has promoted the deployment of all-women UN peacekeeping troops. The UN claims that they have already made an observable difference to peacebuilding outcomes in Liberia where the first all-women troops have been stationed since 2008.[40] While women's capacity for peacekeeping may be practically more nurtured in women-only than mixed groups, such claims about women's more peaceful approach tend to essentialize women's difference from men rather than emphasizing women's equality with men. The UN Secretariat has repeatedly emphasized the proven benefits of having more female peacekeepers, especially in regions where sexual violence has been or still is a serious problem.[41] Yet again gender stereotypes are reinforced by the idea implicit in some gender mainstreaming that only the personal experience and gender identity of police or peacekeepers can increase responsiveness to protect and prevent sexual violence as opposed to a strong political mandate supported by leadership performance indicators and professional training of police and peacekeepers regardless of whether they are male or female.[42] The familiar gendered script wherein the largely male peacekeeping troops are expected to "protect" so-called victimized women and children is only partially disrupted by the presence of female peacekeeping troops. If this model of gender mainstreaming in security is transformationist, it is not articulated as such.

How do we judge the success or failure of gender mainstreaming?

A third critique of gender mainstreaming points to the challenges in monitoring and evaluating the outcomes of different models of mainstreaming for women's empowerment and gender equality. Depending on their institutional mandates, gender has been deployed in various ways by different international organizations: as a synonym for women in the UN Human Rights Commission;[43] as a policy focus on equity and justice in private and public spheres in the Rome Statute of the International Criminal Court;[44] as an acknowledgement of the labor market policy implications of differences between women and men in the EU and the ILO;[45] and as a business case for minimizing these gender differences and their impacts in the World Bank and in APEC.[46] This makes it very challenging to know what the intended positive or negative effects of mainstreaming are and how we might assess them. Policy process rather than outcomes are emphasized by organizations; the World Bank, for instance, considers a gender perspective to be successfully mainstreamed "when it is so routine that it provokes neither conflict nor comment."[47] Yet if it is to be a feminist strategy then we would expect mainstreaming to deliver on the goal of gender equality, which would mean discussing the meanings of gender equality in different global contexts and determining some benchmark indicators for measuring progress toward it.

Complicating this evaluation is the reality that some mainstreaming efforts in international organizations stress the importance of recognizing gender *differences* as in model 2 above rather than gender sameness or male norms as one path to gender equality. For instance, in its publication, *The Other Half of Gender*, the World Bank attempts to bring men and masculine roles back into policymaking, arguing that they have been "displaced" by gender equality efforts. The report highlights how the challenge to traditional masculine roles, such as being the family breadwinner and head of household, undermines men's human capital and results in their social, political, and economic marginalization. It suggests that the gap between normative, hegemonic masculinity and the type of masculinity that poor men are able to achieve is exacerbated by efforts to empower women. Although it critiques gender mainstreaming efforts for relying on and replicating the gendered stereotypes as of "women as victim, men as problem," the study implies that reclaiming traditional masculine roles in development is imperative.[48] In this sense, "men-streaming" development requires women to change rather than men. However, the capitalist economic system that produces these gendered social and economic inequalities in the first place goes unchallenged, and the emphasis is on the material rather than on the ideological basis of gender relations.[49]

In a similar way the UN Secretary-General's recent efforts to recruit male leaders to the cause of eliminating violence against women was intended not to change the gendered power imbalance between women and men but to use men's traditional power to give greater visibility to the violence against women issue (just as celebrities, such as actress Nicole Kidman, have been recruited as UN ambassadors for the cause). The Office of the UN Secretary-General's Special Advisor on Gender Issues (OSAGI) has continually stated that women and men both need to be participants in—and subjects of—gender mainstreaming but has not set a goal of parity or equality between them. The Beijing +5 Resolution in 2000 echoed this new emphasis on the role of men, arguing that "policy-making processes require the partnership of women and men," and that "men and boys should … be actively involved and encouraged in all efforts to achieve the goals of the Platform for Action and its implementation" (paragraph 58).[50] A gender mainstreaming initiative is unlikely to be successful if it includes men and boys but does not challenge gendered inequalities between women and men that are among the causes of violence against women and its justification. Likewise mainstreaming initiatives focusing solely on women and girls in bringing about non-violent change will have limited effects given that men and boys are also affected by this violence and are change agents too, not merely perpetrators of violence.

Other approaches to mainstreaming, however, clearly equate gender equality with sameness in the roles and outcomes for women and men. For example, the focus on mainstreaming gender in international trade and economic policies is largely concerned with increasing the numbers of women employees, entrepreneurs, and exporters fully engaged in the market economy like men. Without any consensus on the meaning of or relationship between gender equality and difference, however, it is hard to judge the impacts of gender mainstreaming. Constructivist International Relations scholars have noted that the most contested norms are the least specific, while the least contested and most widely adopted norms are the most specific, that is, those that are formulated as standardized, technical procedures rather than generalized principles.[51] Where gender mainstreaming has gone beyond the rhetoric in international organization practices it has been implemented as an advanced technique requiring specific methodologies of policy analysis, project management, impact assessment, monitoring, indicators and process measurement. But, in so doing, gender is increasingly treated as a policy input rather than a normative ideal based on either equality or the recognition of difference. Ironically, because budgets are "gender mainstreamed" and therefore combined with the general budgets, there is often meager financial support to implement and monitor the outcomes of the strategy.[52] By contrast with this approach to mainstreaming, Diane Elson argues that in order to realize gender equality and rights for women it needs to move beyond the technocratic practice of creating separate or equal budgets for men and women to become a reflexive feminist practice that uncovers hidden and gendered assumptions in budgets.[53] This type of reflexive practice moves the gender budgeting process from one that employs an "add women and stir" approach to one that allows us to make evaluative claims about gender equality and human rights, and track how they are being progressively realized (or not) through the allocation of resources.

Co-optation by other agendas

A fourth criticism of gender mainstreaming is that it has been readily co-opted as a means to achieving other international agendas such as state security and economic growth only indirectly connected with women's rights or empowerment.[54] This is due to mainstreaming's diffuse interpretation across international organizations revealed by the five models we have set out in this chapter as well as its technical bent. Mainstreaming's process-focused approach aligns well with the World Bank's neoliberal agenda encapsulated in the policy framework "gender equality is smart economics," which seeks to incorporate women in global markets. It also fits neatly into the international security agenda—where mainstreaming checklists to prevent sexual violence and codes of conduct are applied by DPKO officials in headquarters and on the ground in missions. Equality and justice become secondary concerns when mainstreaming is implemented by UN agencies, development and multilateral economic organizations to meet these economic growth and security goals. In this context gender mainstreaming policies are primarily targeted at developing countries where gender inequalities are seen as barriers to these non-gender-based Western goals.

Perhaps the most dramatic example of co-optation is the way in which gender mainstreaming goals have been shrunk to fit the UN Security Council mandate to eliminate sexual violence in armed conflict (for example, in the UN mission in the Democratic Republic of Congo

(MONUC)). This focus seems appropriate given the gravity of sexual violence in some conflict regions but it risks perpetuating the invisibility of various other forms of violence against women, including routine domestic violence in peacetime and within ethnic/national borders. Lacking any thoroughgoing analysis of the complex gendered social, political and economic inequalities that shape women's vulnerability to violence, UN mainstreaming of the sexual violence mandate reinforces gender stereotypes of women as victims and objects of male protection—and men as the power holders.[55] A transformative mainstreaming approach suggested by model 4 would address the causes of egregious sexual violence in conflict/post-conflict settings and involve the provision of social and health services, access to justice and to economic resources rather than merely a state security response.

At the meso-level, NGOs are increasingly responsive to the gender mainstreaming directives linked to non-gender, development and security goals laid out by the World Bank and other international donors. This has resulted in the complex and problematic role shift of many NGOs from being primarily advocates of change to being the primary implementers of development and human rights projects.[56] NGOs pursuing independent projects are often required to implement gender mainstreaming as a condition of funding.[57] Desai's study of gender mainstreaming practices of NGOs in Mumbai reveals increased donor funding for those NGOs that have adopted gender equality, but not necessarily feminist, mandates.[58] Desai's findings also show that many NGOs changed their language from "women" to "gender" and adopted gender mainstreaming language precisely as a way to depoliticize feminist agendas. She concludes that NGOs' increased focus on gender initiatives is not due to the demands of the local populations they serve but is donor driven with much of the funding earmarked for microcredit projects. The Indian government has been supportive of these initiatives since it wants to be seen as complying with the BPA.

Where is the women's movement in gender mainstreaming?

Krook and True argue that gender equality is often left undefined in UN discourses due to the marginality of women's movement voices in debates about gender mainstreaming and the dominance of state actors with contrasting views on the meanings of "gender" and "equality."[59] This point raises a fifth critique of gender mainstreaming by feminists; that it is a bureaucratic process from which women's movements have been largely excluded. For instance, although mainstreaming has deepened and expanded the realm of gender equality policy in the UN system, it is officials who have largely controlled its interpretation and application negating the agency of women's movements in a way that has not occurred with respect to other feminist international strategies such as increasing women's political rights.[60] Some feminist scholars argue that gender mainstreaming has shut down the avenues for civil society influence by bureaucratizing feminism,[61] and attempting to feminize the bureaucrats![62] NGOs have been under pressure to hire more women, regardless of their feminist expertise, as a demonstration of their commitment to gender mainstreaming in the organization. In some cases, such as the largest Cambodian human rights NGO, increasing the number of female employees has actually served to devalue their contributions, which goes against the intent of gender mainstreaming.[63]

Others have argued that gender mainstreaming has been substantially weakened by being both elitist—involving few stakeholders or voices—and subject to technical fads such as mainstreaming guides, training manuals, checklists and toolkits without any substantive issue-based content.[64] Furthermore, the idea that a gender perspective will be made *routine* and therefore *normalized* as opposed to reflective practice runs contrary to feminist struggles to make visible structurally hidden power relations. The arguments behind any feminist strategy intended to progress gender equality always need to be refreshed in the (global) public sphere rather than confined to the administrative sphere. Without participatory debate that allows women and men living under local patriarchal customs to find a way to buy into global gender equality norms, gender mainstreaming may be vulnerable to co-optation, attrition or even reversal as new norms emerge and take over the public and organizational space.[65]

Within World Bank and "UN circuits, feminist contestations are [now] necessarily reactive, responding to priorities, programs, and practices"[66] as determined by others. However, in recent years transnational feminist activists have become more proactive in their demands to have a greater voice and more participatory avenues in global governance. For example, the Gender Equality Architecture Reform (GEAR) Campaign[67] is a coalition of 300 women's, human rights, and social justice civil society organizations which have campaigned for the last five years for the creation of a new UN entity on women and gender equality that would have equivalent standing to other UN agencies, such as the United Nations Children's Fund (UNICEF). As a result of the GEAR Campaign, UN Women was officially created in January 2011.[68] It merges the Division for the Advancement of Women (DAW), International Research and Training Institute for the Advancement of Women (INSTRAW), Office of the Special Advisor on Gender Issues and

Advancement of Women (OSAGI), and United Nations Development Fund for Women (UNIFEM) into one agency. The hope is that by having a consolidated, rather than fragmented, presence, UN Women will be able to be more proactive in progressing gender equality initiatives across the UN and its member states.

Conclusion

In the first part of this chapter we identified five models of gender mainstreaming currently present within international and nongovernmental organizations. Some of these mainstreaming models are both problematic and unrecognizable to many feminists concerned with gender equality and justice.[69] As feminists, we tend to prefer the fourth model, which seeks to transform gender relations within organizations enabling them to serve as catalysts for women's empowerment and for dramatically improving pervasive and globalized gender inequalities. In the second part of the chapter we examined five complex conundrums for feminist researchers in assessing practices of mainstreaming given this transformative intent. Here we consider what is needed to move beyond the shortcomings of mainstreaming and make it deliver as a feminist strategy for global change.

Gender mainstreaming is not passé but central to the new UN Women's mandate. The UN Women's Executive Director Michelle Bachelet, former president of Chile, underscored the importance of gender mainstreaming in her July 2011 speech, which echoes much of our analysis in this chapter:

> Much more remains to be done to mainstream a gender perspective systematically and effectively into all the programmes across the United Nations system. Despite availability of some capacity to carry out the work on gender equality, the focus of this work is mostly on process rather than on impact and results.[70]

Bachelet also highlights the need for a transformative vision of gender mainstreaming that reflects the desires and changes that women want to see. Recognizing the important roles that women's and gender equality NGOs play in global governance, UN Women's mandate calls for increased civil society participation and consultation. To this end, UN Women consulted widely with civil society groups at both the country and regional levels for input into its first Strategic Plan (2011–13).[71] Crucially the GEAR campaign has called for a new approach to governing gender equality involving significant civil society participation in UN Women. This approach would mandate 1) UN staff liaisons with civil society; 2) the creation of civil society councils at all levels; 3) the establishment of thematic working groups; 4) the appointment of independent civil society experts to act as advisors; 5) holding forums for the exchange of ideas with UN Women; and 6) full participation on the governing board of UN Women by civil society representatives.[72] Such initiatives, were they to be adopted, would go a long way to addressing the disconnect between the gender mainstreaming remit and feminist goals. Above all, they would allow UN Women to be a catalyst for the transformative model of gender mainstreaming, challenging hierarchical gender norms and maintaining the achievement of gender equality as a global objective in its own right as well as being a means to greater international security and development.

Notes

1 Mona-Lena Krook and Jacqui True, "Rethinking the Life Cycles of International Norms: The United Nations and the Global Promotion of Gender Equality," *European Journal of International Relations*, 18, 1(2012) 103–128.

2 See BPA, paragraph 202.

3 BPA does not define gender equality but the Office of the Special Advisor to the UN Secretary-General on Gender Issues (OSAGI) has: "Equality between women and men (gender equality): refers to the equal rights, responsibilities and opportunities of women and men and girls and boys. Equality does not mean that women and men will become the same but that women's and men's rights, responsibilities and opportunities will not depend on whether they are born male or female. Gender equality implies that the interests, needs and priorities of both women and men are taken into consideration, recognizing the diversity of different groups of women and men. Gender equality is not a women's issue but should concern and fully engage men as well as women. Equality between women and men is seen both as a human rights issue and as a precondition for, and indicator of, sustainable people-centered development." See UN Women 2012, *Concepts and Definitions*, http://www.un.org/womenwatch/osagi/conceptsandefinitions.htm.

4 OSAGI summarizes this ECOSOC definition: Gender mainstreaming is stated as "the process of assessing the implications for women and men of any planned action, including legislation, policies or programmes, in all areas and at all levels. It is a strategy for making women's as well as men's concerns and experiences an integral dimension of the design, implementation, monitoring and evaluation of policies and programmes in all political, economic and societal spheres so that women and men benefit equally and inequality is not perpetuated. The ultimate goal is to achieve gender equality." Office of the Special Advisor on Gender Issues and the Advancement of Women (OSAGI), *Gender Mainstreaming: An Overview* (New York: United Nations, 2002). http://www.un.org/womenwatch/osagi/pdf/e65237.pdf

5 Jacqui True, "Mainstreaming Gender in Global Public Policy," *International Feminist Journal of Politics* 5, no. 3 (2003): 368–96; Elisabeth Prügl

and Audrey Lustgarten, "Mainstreaming Gender in International Organizations," in *Women and Gender Equity in Development Theory and Practice: Institutions, Resources, and Mobilization*, ed. Jane S. Jaquette and Gale Summerfield (Durham, NC: Duke University Press, 2007), pp. 53–70; Elisabeth Prügl, *The Global Construction of Gender: Home-Based Work in the Political Economy of the 20th Century* (New York: Columbia University Press, 1999).

6 Jacqui True and Michael Mintrom, "Transnational Networks and Policy Diffusion: The Case of Gender Mainstreaming," *International Studies Quarterly* 45, no. 1 (2001): 27–57.

7 Jacqui True, "Gender Mainstreaming and Trade Governance in the Asia-Pacific Economic Cooperation Forum (APEC)," in *Global Governance: Feminist Perspectives*, ed. Shirin Rai and Georgina Waylen (New York: Palgrave, 2008a); Jacqui True, "Global Accountability and Transnational Networks: The Women Leaders Network and the Asia Pacific Economic Forum," *Pacific Review* 21, no. 1 (2008b): 1–26; Jacqui True, "Trading-off Gender Equality for Global Europe: The European Union and Free Trade Agreements," *European Foreign Affairs Review* 14, no. 4 (2009a): 723–42; Elisabeth Prügl, "Does Gender Mainstreaming Work? Feminist Engagements with the German Agricultural State," *International Feminist Journal of Politics* 11, no. 2 (2009): 174–95; Emilie Hafner-Burton and Mark A. Pollack, "Mainstreaming Gender in Global Governance," *European Journal of International Relations* 8, no. 3 (2002): 339–73.

8 Hilary Charlesworth, "Not Waving but Drowning: Gender Mainstreaming and Human Rights in the United Nations," *Harvard Human Rights Journal* 18 (Spring 2005): 1–18; Marie Powell, "A Rights-Based Approach to Gender Equality and Women's Rights," *Canadian Journal of Development Studies* 26, supl (2005): 605–17; Eva Rathgeber, "Gender and Development as a Fugitive Concept," *Canadian Journal of Development Studies* 26, supl (2005): 579–91; Sylvia Walby, "Gender Mainstreaming: Productive Tensions in Theory and Practice," *Social Politics: International Studies in Gender, State, and Society* 12, no. 3 (2005): 321–43; Cynthia Wood, "Different Commonalities: Gender Mainstreaming and the Marginalization of Difference in Economic Development," *Canadian Journal of Development Studies* 26, supl (2005): 593–603; Sally Baden and Anne Marie Goetz, "Who Needs [Sex] When You Can Have [Gender]?," *Feminist Review* 56, no. 1 (1997): 3–25; Rounaq Jahan, *The Elusive Agenda: Mainstreaming Women in Development* (London: Zed Books, 1995); Jacqui True, "Gender Mainstreaming in International Institutions," in *Gender Matters in Global Politics*, ed. Laura J. Shepherd (New York: Routledge, 2010).

9 Nicholas Piálek, "Gender Mainstreaming in Development Organizations: Organizational Discourse and the Perils of Institutional Change," in *Building Feminist Movements and Organizations: Global Perspectives*, ed. Lydia Alpízar Durán, Noël D. Payne and Anahi Russo (London: Zed Books, 2007), 78–86; Rebecca Tiessen, *Everywhere/Nowhere: Gender Mainstreaming in Development Agencies* (Bloomfield, CT: Kumarian Press, 2007); Vandana Desai, "NGOs, Gender Mainstreaming and Urban Poor Communities in Mumbai," in *Mainstreaming Gender in Development: A Critical Review*, ed. Fenella Porter and Caroline Sweetman (Oxford: Oxfam, 2005), 90–98; Senorina Wendoh and Tina Wallace, "Re-thinking Gender Mainstreaming

in African NGOs and Communities," in *Mainstreaming Gender in Development: A Critical Review*, ed. Fenella Porter and Caroline Sweetman (Oxford: Oxfam, 2005), 70–79.

10 Susanne Schech and Mochamad Mustafa, "The Politics of Gender Mainstreaming Poverty Reduction: An Indonesian Case Study," *Social Politics* 17, no. 1 (2010): 111–35; Anne-Marie McGurran, "Gender Mainstreaming and the Public Policy Implementation Process: Round Pegs in Square Holes?" *Policy & Politics* 37, no. 2 (2009): 215–33; Shirin Rai, ed., *Mainstreaming Gender, Democratising the State? Institutional Mechanisms for the Advancement of Women* (Manchester: Manchester University Press, 2003).

11 Emilie M. Hafner-Burton and Mark A. Pollack, "Mainstreaming Gender in the European Union: Getting the Incentives Right," *Comparative European Politics* 7, no. 1 (2009): 114–38; True, "Gender Mainstreaming and Trade Governance in the Asia-Pacific Economic Cooperation Forum (APEC)"; True, "Trading-off Gender Equality for Global Europe: The European Union and Free Trade Agreements," 723–242; Catherine Hoskyns "Governing the EU: Gender and Macroeconomics," in *Global Governance: Feminist Perspectives*, ed. Shirin M. Rai and Georgina Waylen (New York: Palgrave Macmillan, 2008), 107–28; Stefanie Woehl, "Global Governance as Neo-liberal Governmentality: Gender Mainstreaming in the European Employment Strategy," in *Global Governance: Feminist Perspectives*, ed. Shirin M. Rai and Georgina Waylen (New York: Palgrave Macmillan, 2008), 64–83.

12 Laura Parisi, "Reclaiming Spaces of Resistance: Women's Human Rights and Global Restructuring," in *Gender and Global Restructuring: Sightings, Sites and Resistances*, ed. Marianne H. Marchand and Anne Sisson Runyan (New York: Routledge, 2011), 201–22; Prügl and Lustgarten, "Mainstreaming Gender in International Organizations," 53–70; UNDP, *Transforming the Mainstream: Gender in UNDP* (New York: United Nations Development Programme, 2003), http://www.undp.org/women/docs/publication-transforming-the-mainstream.pdf.

13 Fenella Porter and Caroline Sweetman, *Mainstreaming Gender in Development: A Critical Review* (Oxford: Oxfam, 2005); Jahan, *The Elusive Agenda: Mainstreaming Women in Development*.

14 Hafner-Burton and Pollack, "Mainstreaming Gender in the European Union: Getting the Incentives Right," 114–38; Mary Daly, "Gender Mainstreaming in Theory and Practice," *Social Politics: International Studies in Gender, State, and Society* 12, no. 3 (2005): 433–50; Alison Woodward, "European Gender Mainstreaming: Promises and Pitfalls of Transformative Policy," *Review of Policy Research* 20, no. 1 (2003): 65–88.

15 Schech and Mustafa, "The Politics of Gender Mainstreaming Poverty Reduction: An Indonesian Case Study," 111–35; McGurran, "Gender Mainstreaming and the Public Policy Implementation Process: Round Pegs in Square Holes?" 215–33; Ramya Subrahmanian, "Making Sense of Gender in Shifting Institutional Contexts: Some Reflections on Gender Mainstreaming" in *Feminisms in Development: Contradictions, Contestations, and Challenges*, ed. Andrea Cornwall, Elizabeth Harrison and Ann Whitehead (London: Zed Books, 2007), 112–21; Maretha de Waal, "Evaluating Gender Mainstreaming in Development Projects," *Development in Practice* 16, no. 2 (2006): 209–14; Rachael Lorna Johnstone, "Feminist Influence on United

Nations Human Rights Treaty Bodies," *Human Rights Quarterly* 28, 1 (2006): 148–85; Caroline Moser and Annalise Moser, "Gender Mainstreaming since Beijing: A Review of the Success and Limitations in International Institutions," *Gender and Development* 13, no. 2 (2005): 11–22; Aruna Rao and David Kelleher, "Is There Life after Gender Mainstreaming?" *Gender and Development* 13, no. 2 (2005): 57–69.

16 Judith Squires, "Is Mainstreaming Transformative? Theorizing Mainstreaming in the Context of Diversity and Deliberation," *Social Politics* 12, no. 3 (2005): 366–88.

17 Jahan, *The Elusive Agenda: Mainstreaming Women in Development*, 13; Porter and Sweetman, *Mainstreaming Gender in Development: A Critical Review*, 2.

18 Maria C. Correia and Ian Bannon, ed. *The Other Half of Gender: Men's Issues in Development* (Washington, DC: The World Bank, 2006); Desai, "NGOs, Gender Mainstreaming and Urban Poor Communities in Mumbai," 90–98.

19 Jane Parpart, "Fine Words, Failed Policies: Gender Mainstreaming in an Insecure and Unequal World," in *Development in an Insecure and Gendered World: The Relevance of the Millennium Goals*, ed. Jacqueline Leckie (London: Ashgate Publishers, 2009), 60.

20 Prügl, "Does Gender Mainstreaming Work?"

21 Squires, "Is Mainstreaming Transformative?" 375.

22 Emanuela Lombardo and Mieke Verloo, "Institutionalizing Intersectionality in the European Union," *International Feminist Journal of Politics* 11, no. 4 (2009): 478–95; Kimberle Willams Crenshaw, "Gender-Related Aspects of Race Discrimination," United Nations DAW, PHCHR, UNIFEM, Background Paper for the Expert Meeting November 21–24 (Zagreb, 2002).

23 Brooke Ackerly, "Women's Human Rights Activists as Cross-Cultural Theorists," *International Feminist Journal of Politics* 3, no. 3 (2001): 311–46.

24 Rounaq Jahan, *The Elusive Agenda: Mainstreaming Women in Development*, 13.

25 Tiessen, *Everywhere/Nowhere*; Desai, "NGOs, Gender Mainstreaming and Urban Poor Communities in Mumbai," 90–98; Maria Stratigaki, "Gender Mainstreaming versus Positive Action: An On-going Conflict in EU Gender Equality Policy," *European Journal of Women's Studies* 12, no. 2 (2005): 165–86.

26 Two instruments, the gender policy marker and sector code, are used in OECD DAC statistics and allow the OECD to score donors' gender mainstreaming of aid budgets and monitor aid flows targeting gender equality.

27 OECD, *Aid in Support of Gender Equality and Women's Empowerment: Statistics Based on DAC Members Reporting on the Gender Equality Policy Marker 2008–9* (Paris: OECD, 2011).

28 Stratigaki, "Gender Mainstreaming versus Positive Action"; Petra Debusscher and Jacqui True, "Lobbying the EU for Gender-Equal Development," in *The European Union and the Social Dimension of Globalization*, ed. Jan Orbie and Lisa Tortell (New York: Routledge, 2008), 186–206.

29 The OECD Paris Declaration on Aid Effectiveness (2005) is a best practices framework to improve the quality and impact of foreign aid by OECD countries.

30 Maruja Barrig, "What is Justice? Indigenous Women in Andean Development Projects," in *Women and Gender Equity in Development Theory and Practice*, ed. Jane S. Jaquette and Gale Summerfield (Durham, NC: Duke University Press, 2006), 107–34.

31 At the time of writing, 37 National Action Plans on Resolution 1325 have been submitted to the UN Security Council.

32 In the case of developed nations this includes the provision of aid to developing countries to develop and implement their own action plans.

33 Kate Bedford, "Governing Intimacy in the World Bank," in *Global Governance: Feminist Perspectives*, ed. Shirin M. Rai and Georgina Waylen (New York: Palgrave and Macmillan, 2008), 84–106; also Suzanne Bergeron, "The Post-Washington Consensus and Economic Representations of Women in Development at the World Bank," *International Feminist Journal of Politics* 5, no. 3 (2003): 397–419.

34 Jennifer Chan-Tiberghian, "Gender-Skepticism or Gender-Boom? Poststructural Feminisms, Transnational Feminisms, and the World Conference against Racism," *International Feminist Journal of Politics* 6, no. 3 (2004): 454–84; Diane Otto, "Holding up Half the Sky, but for Whose Benefit? A Critical Analysis of the Fourth World Conference on Women," *Australian Feminist Law Journal* 6, no. 1 (1996): 7–28.

35 Lombardo and Verloo, "Institutionalizing Intersectionality in the European Union," 478–95; Squires, "Is Mainstreaming Transformative?" 366–88.

36 UNSC Resolution 1325 (2000) calls on member states to "ensure increased representation of women at all decision-making levels … for the prevention, management, and resolution of conflict" (paragraph 1). It directs the UN Secretary-General to "appoint more women as special representatives and envoys" (paragraph 3) and "expand the role and contribution of women in United Nations field-based operations, and especially among military observers, civilian police, human rights and humanitarian personnel" (paragraph 4).

37 Doreen Carvajal, "The Female Approach to Peacekeeping," *The New York Times*, 5 March 2010, http://www.nytimes.com/2010/03/06/world/africa/06iht-ffpeace.html?ref=global-home.

38 See Jacqui True, "The Unfulfilled Mandate: Gender Mainstreaming and UN Peace Operations," *Georgetown Journal of International Affairs* 10, no. 2 (2009c): 41–51.

39 World Bank, *World Development Report 2011: Conflict, Security, and Development* (Washington, DC: World Bank, 2011).

40 No official evaluation has been conducted although an experimental research design assessing the impact of all-female peacekeeping troops in some regions of Liberia relative to those regions where only all-male or mixed troops are stationed would be feasible to conduct.

41 "Women UN Peacekeepers More Needed—an Advantage." IRIN Global—Humanitarian News and Analysis—a project of the UN Office for Humanitarian Affairs, 20 May 2010.

42 For instance, it has been possible to train police in New Zealand to protect and prosecute domestic violence incidents, which represent over 50 percent of all police call-outs although more than 85 percent of police are male (notwithstanding efforts to promote gender integration within the New Zealand Police Force).

43 Ladma Radmani, "Gender Mainstreaming in the United Nations Human Rights Treaty Bodies," PhD diss., University of Sydney, 2005.

44 Louise Chappell, "The International Criminal Court: A New Arena for Transforming Gender Justice," in *Global Governance: Feminist Perspectives,*

ed. Shirin M. Rai and Georgina Waylen (New York: Macmillan Palgrave, 2008), 160–84.

45 True, "Trading-off Gender Equality for Global Europe," 723–242; Prügl and Lustgarten, "Mainstreaming Gender in International Organizations."

46 Bedford, "Governing Intimacy in the World Bank"; Bergeron, "The Post-Washington Consensus and Economic Representations of Women in Development"; True, "Global Accountability and Transnational Networks," 1–26.

47 Bedford, "Governing Intimacy in the World Bank," 95–96, quoting World Bank.

48 Correia and Bannon, ed., *The Other Half of Gender*, 253.

49 Laura Parisi, "'Disciplining' and 'Engendering' the World Bank: A Comment," in *Feminist Economics and the World Bank: History, Theory, and Policy*, ed. Edith Kuiper and Drucilla K. Barker (New York: Routledge, 2006), 197–206.

50 Krook and True, "Rethinking the Life Cycles of International Norms."

51 Antje Wiener, "Enacting Meaning-in-Use: Qualitative Research on Norms and International Relations," *Review of International Studies* 35, 1 (2009): 185.

52 Debusscher and True, "Lobbying the EU for Gender-Equal Development," 186–206; AWID (Association for Women in International Development), *Where's the Money in Women's Rights: Assessing the Resources and the Role of Donors in the Promotion of Women's Rights and the Support of Women's Rights Organizations* (First Fundher Report, 2006).

53 Diane Elson, *Budgeting for Women's Rights; Monitoring Government Budgets for Compliance with CEDAW* (New York: UNIFEM, 2006). For Elson, spending should equally serve the needs and priorities of men and women with the goal of reducing gender inequality. To achieve this end, she suggests the following questions must be asked: "What impact does this fiscal measure have on gender equality? Does it reduce it; increase it; or leave it unchanged?" (Diane Elson, "Gender Mainstreaming and Gender Budgeting," paper presented at Gender and Equality and Europe's Future Conference, 4 March 2003, 3, http://ec.europa.eu/education/programmes/llp/jm/more/conf gender03/elson.pdf).

54 Parisi, "Reclaiming Spaces of Resistance," 209–13; Jacqui True, "Feminist Problems with International Norms: Gender Mainstreaming in Global Governance," in *Feminism and International Relations: Conversations about the Past, Present and Future*, ed. J. Ann Tickner and Laura Sjoberg (New York: Routledge, 2011), 73–89.

55 In Resolutions 1325, 1820, 1888 and 1889 the UN Security Council has linked the scourge of sexual violence to international insecurity and stressed that it needs to be addressed in all conflict situations. In Resolution 1888, ending sexual violence in conflict is elevated above other forms of violence against women and all other issues and objectives relating to women and gender equality in peace and security, including the importance of women's participation in conflict-resolution and peacebuilding; see True, "Gender Mainstreaming in International Institutions."

56 Sangeeta Kamat, "The Privatization of Public Interest: Theorizing NGO Discourse in a Neoliberal Era," *Review of International Political Economy* 11, no. 1 (2004): 155–76. For example, the participation by civil society organizations, which includes NGOs, in World Bank projects has increased from 21 percent in 1990 to 72 percent in 2005. World Bank, "10 Things

You Never Knew About the World Bank: #8" (n.d.). Available at: http://digitalmedia.worldbank.org/tenthings/en/8.php.

57 Tiessen, *Everywhere/Nowhere*; Wendoh and Wallace, "Re-thinking Gender Mainstreaming in African NGOs and Communities," 73; Desai, "NGOs, Gender Mainstreaming and Urban Poor Communities in Mumbai," 95–96.

58 Desai, "NGOs, Gender Mainstreaming and Urban Poor Communities in Mumbai," 95.

59 Krook and True, "Rethinking the Life Cycles of International Norms."

60 This is not withstanding that many UN officials in gender equality policy are "insider-outsiders" with connections to women's movements outside the UN.

61 See on "NGOisation": Sonia Alvarez, "Beyond NGOization: Reflections from Latin-America," *Development* 52, no. 2 (2009): 175–84.

62 Jacqui True, "Gender Specialists and Global Governance Organizations," in *Women's Movements: In Abeyance or Flourishing in New Ways?* ed. Marian Sawer and Sandra Grey (New York: Routledge, 2009b), 91–104.

63 Maitrayee Mukhopadhyay, "Mainstreaming Gender or 'Streaming' Gender Away: Feminists Marooned in the Development Business," in *Feminisms in Development: Contradictions, Contestations, and Challenges*, ed. Andrea Cornwall, Elizabeth Harrison and Ann Whitehead (London: Zed Books, 2007), 135–49.

64 Mieke Verloo, "Displacement and Empowerment: Reflections on the Concept and Practice of the Council of Europe Approach to Gender Mainstreaming and Gender Equality," *Social Politics: International Studies in Gender, State and Society* 12, no. 1 (2005): 344–65.

65 Krook and True, "Rethinking the Life Cycles of International Norms."

66 Mary E. Hawkesworth, *Globalization and Feminist Activism* (Lanham, MD: Rowman & Littlefield, 2006): 143.

67 See GEAR 2012, *Home*, http://www.gearcampaign.org/ for more information.

68 See UN Women 2012, *UN Women, United Nations Entity for Gender Equality and the Empowerment of Women*, http://www.unwomen.org/.

69 Ines Smyth, "Talking of Gender: Words and Meanings in Development Organisations," *Development in Practice* 17, no. 4–5 (2007): 548.

70 UN Women 2012, *Mainstreaming a Gender Perspective into UN Policies and Programmes*, http://www.unwomen.org/2011/07/mainstreaming-a-gender-perspective-into-un-policies-and-programmes/.

71 See GEAR 2010, *Civil Society Participation and UN Women*, http://www.gearcampaign.org/news_events/wp-content/uploads/2010/11/November-2010-GEAR-Civil-Society-Participation-brief.pdf.

72 See GEAR 2011, *Powerpoint on UN Women and GEAR*, http://www.gearcampaign.org/news_events/uncategorized/powerpoint-on-un-women-and-gear/.

3 Gender expertise as feminist strategy[1]

Elisabeth Prügl

- **Data and methodology**
- **Ends, logics, and techniques**
- **Conclusion**

One of the remarkable outcomes of the feminist movement of the late twentieth century has been the development of specialized knowledge about gender relations and the packaging of this knowledge as expertise. The development of gender expertise exploded in response to the wide adoption of gender mainstreaming in international organizations (IOs) and governments. Feminists, organizational consultants and trainers met the demand for gender training, gender analysis, and organizational change, in part bringing to bear their own background in gender studies, in part adding gender expertise to their existing qualifications. Thus, there exists today an international cadre of gender experts who play a key role in translating feminist knowledge into policy applications. UN Women epitomizes such expertise in its professional staff and because it provides an organizational focus for gender experts. What is the significance of these gender experts in advancing feminist agendas? To what extent do they and their expertise constitute feminist strategies?

The emergence of gender experts has generated unease among some feminists who have feared a stifling of radical feminist movement ideas and the co-optation of feminist agendas for governmental purposes.[2] But expertise also has long been recognized for its contribution to solving complex problems and its ability to rationalize governance at the international level.[3] Scientists have long participated in analyzing problems ranging from underdevelopment to environmental degradation and their knowledge is considered crucial to developing solutions that work. Similarly, gender expertise can be considered as a strategic tool to help pinpoint the reasons for continued gender inequality and subordination, and to find ways to advance feminist goals.

With the adoption of a number of Security Council resolutions on gender, peace, and security in the new century, gender expertise has gained an unprecedented salience in the security sector. The resolutions have defined a three-fold agenda of protection, participation and prevention. First they call on governments and the United Nations (UN) to provide women protection in particular from sexual violence during times of war and in processes of post-war reconstruction. Second, they appeal to parties to draw on the skills of women and include them in conflict resolution and in decision-making in the rebuilding of societies. And finally, recognizing that gender inequality and gender-based violence contribute to the outbreak of war and to sexual violence during war, they urge governments and UN agencies to mainstream gender considerations into early warning mechanisms and into the training of militaries and peacekeepers. While the women, peace and security agenda is thus fairly broad, in practice governments, UN agencies, and non-governmental organizations (NGOs) have focused disproportionately on the problem of gender-based violence and largely disregarded a focus on how considerations of gender could be inserted into efforts to prevent wars.

The mandate of inserting considerations of gender into peace and security activities has led to a proliferation of capacity building activities in the form of gender training. Gender experts are being recruited to conduct such training and to develop training materials. In this paper I explore these materials for what they tell us about the strategic quality of gender expertise in the security sector. Gender expertise is feminist strategy because it is being deployed as a tool to advance feminist goals—however vaguely these may be defined. As outlined in the introduction to this book, such strategy cannot be understood merely as a matter of relating means to ends, of applying the best tools in order to accomplish desired goals. Instead, the means and instruments strategically deployed with an eye towards feminist ends invariably develop their own logics. Accordingly I ask, how does feminist expertise operate? What ends does it define, what arguments does it advance in support of these ends, how does it gain authority, and what are the power effects of such authority? I draw on a strand of Foucaultian-inspired feminist writings that probes the kinds of politics gender expertise makes possible and the kinds of options it precludes.[4] The purpose of this approach is not an unveiling of inadvertent ideology or a measuring of practice against an abstract standard of true feminism. Instead, it is to make visible the power politics that is inherent in any strategic effort to change the world.

Data and methodology

Gender training manuals are produced by gender experts in IOs and NGOs, and by subcontractors in the private sector. I provide a close reading of 12 such manuals in the security sector. They represent a sample of the 52 manuals, tool kits, and background papers that were listed under "Gender Peace and Security" on the UN International Research and Training Institute for the Advancement of Women (INSTRAW) website in the spring of 2010 and were associated with the INSTRAW Community of Practice on Gender Mainstreaming.[5] The 12 manuals selected are listed in the annex. The sample represents a balance between manuals produced or commissioned by NGOs, IOs, and governments. Of the 12 manuals, four are part of a package produced by the Geneva Center for the Democratic Control of Armed Forces (DCAF) and two are written by International Alert together with Women Waging Peace (IA/WWP) as part of a series. So arguably I am dealing with eight cases.

Of these eight cases, three are manuals produced or commissioned by UN agencies, two by NGOs, two under subcontract from the US government, and one by a quasi-governmental organization. Of the IO manuals, two target humanitarian workers, one in the area of disarmament, demobilization, and reintegration (UN Development Fund for Women—UNIFEM) and one on the topic of gender-based violence (UN International Children's Emergency Fund). One UNIFEM manual addresses itself to advocates interested in arguing gender equality in the security sector with the help of the Convention on the Elimination of All Forms of Discrimination Against Women (CEDAW). Of the two cases of NGO manuals one also targets peace activists and seeks to sensitize them to the role of women in fighting militarism and war (IA/WWP). Another, produced by Reach Out, a refugee protection training project initiated in 2001 by NGOs and the Red Cross/Red Crescent Movement, aims to sensitize humanitarian workers on the issue of gender-based violence. Both of the manuals financed by the US government address the issue of gender-based violence, one seeking to enlist the help of religious leaders, the other addressing itself to humanitarian personnel.[6] Finally, DCAF wrote a package of manuals and tool kits in support of security sector reform, a selection of which is included in my analysis. DCAF is an international foundation funded by the Swiss government to advance security sector governance.

While there are differences between the manuals based on the entities that authored them, their similarities are perhaps more striking. Gender-based violence is a preoccupation for UN agencies, NGOs, and the US government alike. Similarly, both NGOs and UNIFEM are interested in training advocates and activists. This pattern of overlapping interests and targets supports a Foucaultian interpretation, which suggests that these organizations all are part of an overall practice of governmentality; they come together in the shared purpose of governing the conduct of individuals in such a way that they advance a particular understanding of the common good, i.e., "security" that does not entail women-despising practices. Experts generate the knowledge necessary for this practice. They could be considered the Trojan horses of feminism in governmental processes and in this sense a rational strategic instrument. But they also reinterpret feminism to match governmental processes and thus become the Trojan horses of governmentality in feminism. In this sense, gender experts and gender training are instruments in the government of gender.

At the center of expert strategizing is the construction of knowledge, framing issues and providing information in a fashion that convinces. Such knowledge is power not only in the sense that it gives individuals or groups an advantage in argumentation, but also in the sense that it makes various forms of conduct thinkable. It defines particular identities, fixes categories of things, validates ideals, and introduces logics that drive towards the creation of such ideals. Knowledge is power in this sense because it generates realities. These realities may encompass those intended by feminist strategists, but they may encompass in addition identifications and logics that insinuate themselves inadvertently through the adoption of a language of governmentality. Drawing on a Foucaultian analytics of government thus leads me to ask not only what are the ends of government, but also, what identities are being constructed, and what logics and techniques are used in governing.[7]

Ends, logics, and techniques

The gender manuals under consideration can be categorized into three groups depending on the strategic ends that they pursue: some seek to transform security practices, some to achieve gender equality, and some to fight gender-based violence. The set of toolkits produced by DCAF and the two manuals by IA/WWP formulate as their goal the transformation of security. In these manuals addressing gender equality and gender-based violence are issues linked to the larger struggle against militarism and war. Two UNIFEM manuals in contrast see as their end gender equality. Fighting gender-based violence is integral to this end, but the security issue is just one more arena in which the larger struggle for gender equality needs to be fought. Combating gender-based violence is the main aim in the remaining manuals. They typically frame ending such violence as an end in itself, a matter of achieving a basic human right. They recognize gender equality as a major facilitator of this aim. But the fact that their struggle targets the security sector appears largely incidental; they do not tend to connect gender-based violence with militarism and war.

Different logics inhere these manuals and they envision the employment of different techniques of governance, matching the strategic ends they seek to achieve. The logics, their associated ends and techniques are summarized in Table 3.1. The table guides the discussion in the remainder of this section.

Transforming security

The first logic of governmentality deployed in the manuals draws on feminine difference to argue the inclusion of women into security practices. The logic develops in two different directions: on the one hand it envisions a validation of women's difference in security practices and as a result a profound change of such practices. On the other hand it combines this idea with a logic of efficiency to suggest that validating women's difference would perfect existing practices and achieve an operational transformation; it implicitly advocates a cooptation of femininity into such practices.

Essentialist depictions of women are rampant in the manuals that make a transformation of security their major end. The associations

Table 3.1 Training manuals in the security sector: ends, logics, and techniques

End	*Logic*	*Technique*
Transforming security	Gender difference: Operational effectiveness	Story telling: Exemplary cases
Gender equality	Undoing gender: Justice and democracy	Technocratic administration: Reference to CEDAW, Security Council Resolution 1325 (SC 1325) and exemplary cases
Ending gender-based violence	Women are human: Human rights and faith	Technocratic administration: Reference to human rights instruments, commission reports, policies, codes of conduct and statistics Pastoral power: Reference to scriptures

produced are quite varied ranging from notions of women as care givers occupying the private realm to notions of women as peace activists. In these roles, women are described as having unique skills that lend themselves to achieving peace. For example, the manuals jointly produced by International Alert and Women Waging Peace suggest a number of roles that women take, carving out a particular space for them in the private and the public spheres. The manuals associate peacemaking and peacebuilding with care-giving, nurturing and healing. Thus, we learn that

> in conflicts and war-torn countries across the world, women are often at the forefront of peacemaking and peacebuilding efforts. They are leading households, caring for the sick and the old and sustaining and ensuring the survival of their families and communities. Even in refugee camps … women shoulder the responsibilities of others. In many instances … women initiate peacemaking efforts. It is often women who are the mobilisers and the voice of civilians silenced by violence and atrocities. Long after international aid and support have ended, women are left to handle the trauma and the violence that comes home with men … and to provide the continuity that enables families and communities to heal and move forward.[8]

Women in this excerpt are connected to peace, households, caring, civilians, home, families, and communities. They are stereotyped according to imaginary traditional roles.

But, unlike in patriarchal renderings of such narratives, women in the language of the gender experts are not stereotypically passive. On the contrary: They take responsibility, empowered by their particular social positions and having developed unique skills as a result of their specific experiences. Thus, the writers of the IA/WWP manuals suggest that

- women bring an understanding of "the root causes of the conflict";
- women speak effectively "about the impact of violence on daily life, relating the experiences and voices of ordinary people";
- women "tend to have a holistic approach to resolving conflict, often motivated by the dream of a peaceful future, particularly for their children";
- women "tend to focus on practical issues related to quality of life and human security, rather than control over political power"; and
- women recognize "the importance of building positive relations."[9]

The manual goes on to suggest that women are likely to enhance the process of peace negotiations because they are inclusive, consultative, empathic, and trustworthy. It cites psychologist Simon Baron-Cohen who in his 2003 book, *The Essential Difference*, shows "that women and girls tend to be better at empathizing and communicating than their male counterparts."[10]

The logic put forward in these manuals is that women's difference should be made use of in order to transform security practices. And peacemaking is the arena where this can and should be accomplished. The manuals summon individual women to identify as peacemakers and activists to recognize themselves in this role. Allowing women access to peace activism and making their distinctive roles visible in such activism thus promises more success in the struggles for peace and security.

A slightly different, but related logic emerges in those manuals (and parts of manuals) that are concerned less with peacemaking than with peace-support operations. Here women's different aptitudes are put forward to suggest that they can advance operational effectiveness. For example, we are told that women police officers have the following positive attributes. They

- use force less frequently than their male counterparts;
- are less authoritarian when interacting with citizens and lower-ranking officers;
- have better communication and negotiation skills;
- are more likely than male officers to diffuse potentially violent situations;
- respond more effectively to violence committed against women; and
- have significantly lower rates of complaints of conduct, improper use of force or inappropriate use of weapons.[11]

Women also are said to be better able to "gain the trust of civilians, engage and work towards the full involvement of local women, [and] exercise communication and crowd control skills."[12]

In addition to these presumably universal skills and aptitudes, women are needed because their femininity allows particular forms of conduct not considered feasible for men but highly useful to some security operations. For example, the DCAF manual on defense reform suggests that "In certain contexts, the inclusion of women in defense forces is an operational imperative. As well as tasks that both men and women perform, women can carry out critical tasks that men can only take on with difficulty, if at all."[13] These tasks include the screening of female ex-combatants and searching women more broadly, the widening of a net of intelligence gathering, and assisting in the aftermath of sexual violence.

Thus, the logic of women's difference apparently lends itself to varied security ends. On the one hand, women's perceived aptitudes and skills generate a rationale of conflict transformation. In this logic, essentialism becomes strategic because it enables change by defining femininity as a force for peace. On the other hand, women's perceived aptitudes and skills generate a rationale of inclusion for the purposes of operational effectiveness. In this logic essentialism becomes a strategy for military efficiency, implicitly advocating co-optation. In both instances women's difference functions as a generative principle strategically hitched to governmental purposes.

In order to become authoritative, gender manuals employ a number of different techniques. The most important of these is the reference to international law, codes of conducts, and best practices. Among the three types of manuals analyzed here, those aiming for a transformation of security are least able to draw on international law. They point to some relevant injunctions in human rights instruments and UN policies, but ultimately have to go beyond legal texts to make their case. Restricted to logics of effectiveness, their main technique is the presentation of stories of success and failure resulting from the integration of women into peacemaking and the mainstreaming of gender into security operations. We learn about women protesting violence in the Niger delta, the Mano River Women's Peace Network struggling for regional peace in Sierra Leone, Liberia, and Guinea, and women in Northern Ireland advocating for inclusiveness in the peace negotiations.[14] We also learn about the Jamaican effort to create a participatory and gender-responsive national security policy, defense reform in South Africa, and the successes garnered from increasing the number of women in the Hungarian armed forces (DCAF manuals). Narrative evidence grounds arguments about the effectiveness of deploying femininity for the purpose of achieving conflict transformation or operational success proposing truth by practical example. Such evidence invites activists, militaries, and humanitarian agencies to learn from the truths in the stories and imitate the successes described. They summon women to activate their capacities to bring about peace and invite militaries and humanitarian agencies to increase their effectiveness by including women. Gender expertise thus insinuates itself into the construction of individual and collective identities, seeking to guide conduct from a distance while governing security practices with the help of a new apparatus of gender.

Achieving gender equality

A different logic of governmentality is deployed in the UNIFEM manuals that identify gender equality as the end to be achieved. In the words of one UNIFEM manual, peacemaking is "an opportunity to make advancements on women's issues and rights. For women, the peace table becomes a platform for addressing social injustices against women."[15]

These manuals use gender not to make an argument about difference, but seek to "undo gender,"[16] thereby forgoing difference as a generative principle. In particular, the UNIFEM manual on disarmament, demobilization, and reintegration (DDR) refuses an endorsement of women's difference; it adduces instead the notion of gender divisions of labor, and warns that the exclusion of women from political processes "normalizes and reinforces gender roles that associate men with guns, violence and formal politics, and women with passivity and domesticity."[17] The manual illustrates the diverse roles that women play during conflict. It emphasizes that women are not just a vulnerable group and victims, but also combatants and quite willing to commit atrocities. At the same time, the manual emphasizes that women have "different needs" resulting from their different insertion in society and calls on humanitarians to recognize these needs so that women can become equal citizens. Women are reconstructed as citizen soldiers with particular needs.

Gender difference then no longer serves as an argument for women's inclusion—it is not their unique qualities that should be brought to fruition. Instead, in these manuals, the logic of gender difference is replaced with a logic of justice. If both women and men participate in conflict in a diverse range of roles, women should receive the same resources and access that men receive as a matter of fairness. Humanitarians are summoned by the values of liberal democracy: justice and equality. Once women are redefined as citizens, these notions become forceful principles reorganizing worldviews and driving new behaviors. Enhancing women's inclusion and participation becomes enmeshed in a diffusion of liberal values.

In order to assert the authority of their claims, these manuals can add legal arguments to arguments of effectiveness. CEDAW and Security Council Resolution 1325 support the claims in the first UNIFEM manual. Indeed, the explicit purpose of this manual is to teach advocates how to argue with CEDAW and SC 1325 to advance gender equality while building peace and putting in place the lasting conditions for preventing conflict. But the manuals also draw on narrative evidence and case studies. In the second UNIFEM manual (on DDR), two case studies drive the argument, illustrating successes and failures in Liberia and in Bougainville, New Guinea. The cases prepare the introduction of UNIFEM's "standard operating procedures for gender-aware DDR," 10 pages of detailed prescriptions on how to promote women's political participation and inclusion in DDR processes and on how to address their different needs. The cases become blueprints for successes to be imitated and failures to be avoided.

In the expert knowledge displayed in these manuals the rationality of liberal democracy drives a strategy of achieving women's equality, combined with a promise that such equality transforms society so that conflict is prevented in the future. This knowledge draws on the authority of legal texts prescribing equal treatment (CEDAW), women's inclusion, and a consideration of their particular experiences (SC 1325). The language of rights and justice provides the generative logic for the arguments, and this language is translated into the practices of DDR through standard operating procedures. Women are constructed as equal citizens with equal rights, including rights to access resources and participate in political processes.

Combating gender-based violence

Like the UNIFEM manuals, manuals geared towards combating gender-based violence reject images of women's difference. But neither do they evoke women's citizenship. The power of their strategy instead derives from defining women as humans, that is from emphasizing that the category of the human in "human rights" applies to women and that ignoring violence against women amounts to denying them their basic humanity.[18] Three of the four manuals in this category make human rights the focus of their argument; the language of human rights provides them their generative logic. The approach is typified by the understanding that "nobody has to give these rights to you because you have them automatically from birth."[19] They are virtually self-executing. The fourth manual takes a different approach. Targeted towards enlisting religious leaders, it makes no reference to human rights, but extensively cites religious texts. For the most part, these texts appeal to individual responsibility and doing the right thing, to helping the least well-off and caring for the poor, to women and men being respectful towards each other. All the manuals activate pastoral power in the Foucaultian sense, i.e., they invoke adherents to follow the injunctions of a basic set of principles. Spiritual leaders are called upon to guide their flocks by providing appropriate counseling in addition to starting prevention and support programs. Human rights defenders of various stripes are called upon to spread the understanding of peoples' common humanity, of everybody's innate basic rights that should internally drive all actions towards others.

All four manuals assert that women are particularly vulnerable to gender-based violence, but immediately qualify this statement by specifying that men can be victims as well. The issue is not so much one of gender, but of recognizing everybody's humanity. The manuals also follow the convention that has developed among those combating violence against women to call victims "survivors." Gender expertise here seeks to counteract the tendency of constructing victims as passive objects of others' actions, lacking agency and their own designs. Yet the extensive reliance on legal mechanisms and formal enforcement foils this good intention since these technologies invariably approach victims as passive objects to whom things were done by others and to whom justice needs to be provided by legal experts who intervene.

Manuals on gender-based violence are in a privileged position compared to other manuals in that they gain their authority from a variety of techniques. Unlike these other manuals, they do not resort to exemplary case studies and arguments of efficiency. Instead they have at their disposal a large tool kit of human rights instruments, including CEDAW, the Declaration on the Elimination of Violence Against Women, the Convention on the Rights of the Child, the Rome Statute of the International Criminal Court, the UN Protocol on Trafficking, and the various Security Council resolutions, all of which bolster their case. In addition, they can make reference to various policy statements and commission reports. Thus, the UNICEF manual introduces the report of the Inter-Agency Standing Committee Task Force on the Protection from Sexual Exploitation and Abuse in Humanitarian Crises and the related policy statement.[20] Manuals on gender-based violence furthermore can claim the force of codes of conduct established in the UN system and associated threats of sanctions, and they can draw authority from religious texts, as in the United States Agency for International Development (USAID) manual. Unlike any of the other manuals, they also put forward statistical evidence documenting the frequency and scope of gender-based violence. The manuals on gender-based violence thus are able to combine authority derived from legal, policy, and religious texts with quantitative evidence.

These various instruments constitute technologies for the exercise of both pastoral and disciplinary power, that is they shepherd individuals towards the right behavior, and they bring to bear knowledge about lawful and unlawful behavior to generate self-sanctioning subjects. Conventions, religious texts and political statements provide guidance for the conduct of humanitarians and others involved in conflict, and codes of conduct—often drafted by gender experts—threaten punishment if rules are violated. Gender expertise activates the power of "third parties," from heavenly judgment to tribunals, to motivate a change in individual conduct. It is helped in this task by the application of technologies ranging from counseling and confession to correction, achieving amends and transformation after violation. The logic of these technologies invariably targets the perpetrator as the active agent while keeping the victim passive. It summons him (mostly) to adjust his worldview and behavior to a new understanding of women as humans.

In sum, the logics of human rights and religious values of love and respect fuel a strategy of abolishing gender-based violence. The logics refuse gender as a definer of difference, and instead evoke an abstract and feminized vulnerable human, largely deprived of agency. The legality of human rights, and their resonance with basic religious values, provides a powerful technology to deploy this logic and produce realities in contexts of international intervention.

Conclusion

The purpose of this chapter has been to explore the strategic quality of gender expertise. What logics does such expertise develop once deployed in organizational contexts? How does it gain authority and what power effects does it produce? I have identified three different logics corresponding to the strategic ends imputed in security sector training manuals. The first set of manuals seeks a transformation of security practices by linking them to gender; it draws on illustrative case studies to invoke the power of feminine difference and suggest the importance of including women and considering gender relations in planning for peace. In deploying notions of feminine difference, the argument generates identity effects that normalize what it means to be a woman and implicitly exclude those who do not conform, while at the same time advancing the strategic goal of an engendered peace. The second set of manuals defines gender equality as its strategic end; these manuals are more attentive to the power effects of gender categorizations and identity productions, and replace the notion of women's intrinsic difference with the notion of women as rights-bearing citizens. They set in motion liberal arguments of justice to advance the case for women's rights to political participation and for recognition of their different needs. Finally, the manuals seeking to combat gender-based violence draw on human rights instruments and religious texts to activate notions of personal responsibility and a common humanity that women and men share. The law and religion put into operation techniques of counseling in addition to confession and legal intervention that guide, discipline and strategically direct the conduct of populations towards an end of gender-based violence.

The knowledge of gender experts displayed in these manuals thus becomes strategic towards a variety of ends—most of which feminists would endorse. Yet as gender expertise becomes strategic, and as it makes use of technologies that convey its authority, it also develops logics that exercise power in various ways, generating both empowerments and disempowerments. Foremost among these are identity effects, which range from essentialism to displacement undermining the stability of binary gender constructions. They encompass the normalization of what it means to be a woman in a conflict situation in the manuals that seek a transformation of security, the denial of women's agency in the manuals targeting gender-based violence, but also the undoing of gender in the UNIFEM manuals. Related to these are identity effects on humanitarian personnel and on men, i.e., the affirmation of masculinity as militaristic, the disciplining and the shepherding of male sexual and violent inclinations, as well as the refusal to construct women as the other of men with its implied attack on man as the political center.

Gender expertise also exercises power through the technologies it deploys. The most salient technology for inserting gender into discourses of security appears to be human rights law, together with a variety of administrative tools such as codes of conduct. Manuals that seek to end gender-based violence are particularly apt to draw on strong human rights norms to guide conduct through the threat of sanctions, flanked by pastoral technologies appealing to personal responsibility and doing good. These methods carry disciplinary power in that they produce new types of (male) individuals internally guided to do the right thing and abstain from certain kinds of sexual conduct. They function to re-describe a relationship between the humanitarian and the native that virtually precludes sexual encounters. Furthermore, in relegating victims to a passive position and defining them in need of protection, they become unreflexive techniques insensitive to the way in which they hem in identities. Indeed, it appears that expertise that can draw extensively on legal authority is particularly prone to technocracy—the development and application of standard operating procedures, the categorization and cataloging of populations according to statistical patterns together with an objectification of those it intends to help—and particularly blind to the politics of gender-based violence and the need for feminized populations to participate in fighting it.

Expertise that connects gender equality to the goal of transforming conflicts and making peace lacks this kind of the epistemic authority. It advances narrative case studies and best practices to argue feminine difference and suggests that including women increases operational effectiveness. It lacks authoritative technologies to produce corresponding conduct in relevant populations. But this expertise appeals to feminist activism—arguing the need for inclusiveness, forming partnerships and awareness raising.

Expertise that defines the end as gender equality falls between the extremes of a technocratic administration of populations and activating such feminist politics. Like the expertise on gender-based violence, it marshals the legal technologies of human rights law, Security Council resolutions, and best practices. Like the expertise on transforming security by engendering it, it seeks to put these technologies in the service of a new politics that allows for women's participation. UN Women inherits this approach from UNIFEM, which has laid the ground works for a politics that enhances feminist authority while avoiding the dangers of essentializing or victimizing women. The approach carefully combines the effectiveness of legal and administrative technologies to combat gender-based violence with the emancipatory discourse of women's citizenship.

A power analysis of gender training in the security sector thus reveals both opportunities and dangers as gender expertise is strategically deployed. The dangers lie in turning feminist knowledge into technocratic administration—most visible currently in training on gender-based violence—and in an essentializing of women. The opportunities lie in enhancing epistemic authority for feminist ends. The conversion of feminist knowledge into gender expertise cannot be dismissed as a sign of a failed feminist revolution. In a global society increasingly structured by expertise, it constitutes a process whereby feminist knowledge gains authority, unfolding rationalities of government while also battling mechanisms of power and disempowerment.

Annex: manuals analyzed

UN agencies

1. UNIFEM, *Getting it Right, Doing it Right: Gender and Disarmament, Demobilization and Reintegration.* 2004. (Written by Sarah Douglas, Vanessa Farr, Felicity Hill, and Wenny Kasuma.)
2. UNIFEM, *Advancing Gender Equality Using CEDAW and UN Security Council Resolution 1325: Training Module for Gender Equality Advocates.* (Developed by the UNIFEM Regional Project "Women for Conflict Prevention and Peace-Building in the Southern Caucasus" starting in 2003.)
3. UNICEF, *Training of Trainers of Gender-Based Violence: Focusing on Sexual Exploitation and Abuse.* (Draft tested in Nepal in April 2003.)

NGOs

4. International Alert and Women Waging Peace, *Inclusive Security, Sustainable Peace: A Toolkit for Advocacy and Action. Introduction.* Hunt Alternatives Fund and International Alert 2004a. (Written by Adrian-Paul Ancil, Sanam Naraghi Anderlini, Camille Pampell Conaway, Judy El-Bushra, Kelly Fish, Nicola Johnston, Lisa Kays, Gillian Lobo, Elizabeth Powley, Jolynn Shoemaker, Victoria Stanski, Guillermo Suarez Sebastian, and Mebrak Tareke.)

   International Alert and Women Waging Peace, *Inclusive Security, Sustainable Peace: A Toolkit for Advocacy and Action. Conflict Prevention, Resolution and Reconstruction.* Hunt Alternatives Fund and International Alert 2004b. (Written by Sanam Naraghi Anderlini, Victoria Stanski, Nicola Johnston, and Judy El-Bushra.)
5. Reach Out Refugee Protection Training Project, *Optional Module: Gender-based Violence (GBV).* 2005.

Governments

6. JSI Research and Training Institute/RHRC Consortium, *Training Manual Facilitator's Guide: Multisectoral and Interagency Prevention and Response to Gender-based Violence in Populations Affected by Armed Conflict.* 2004. (Written by Beth Vann and funded by the US Department of State.)
7. USAID and Health Policy Initiative, *A Call to Act: Engaging Religious Leaders and Communities in Addressing Gender-Based Violence and HIV.* 2009. (Written by Britt Herstad, Health Policy Initiative.)

Quasi-governmental organization

8. Geneva Centre for the Democratic Control of Armed Forces (DCAF), *Gender and Security Sector Reform Training Resource Package.*
   - *Guide to Integrating Gender in SSR Training.* 2009a. (Written by Kristin Valasek and Agneta M. Johannsen.)
   - *Training Resources on Security Sector Reform and Gender.* 2009b. (Written by Agneta M. Johannsen.)
   - *Training Resources on National Security Policy-Making and Gender.* 2009c. (Written by Agneta M. Johannsen.)
   - *Training Resources on Defence Reform and Gender.* 2009d. (Written by Agneta M. Johannsen.)

Notes

1 Thanks to Nell Williams for her superior research assistance in the analysis of gender training manuals.

2 Angelika Wetterer, "Strategien rhetorischer Modernisierung: Gender Mainstreaming, Managing Diversity und die Professionalisierung der Gender-Expertinnen," *Zeitschrift für Frauenforschung & Geschlechterstudien* 3 (2002): 129–48; Heike Kahlert, "Beratung zur Emanzipation? Gender Mainstreaming unter dem Vorzeichen von New Public Management, " in *Was bewirkt Gender Mainstreaming? Evaluierung durch Policy-Analysen*, ed. Ute Behning and Birgit Sauer (Frankfurt, Germany: Campus, 2005), 45–62.

3 Peter Haas, "Introduction: Epistemic Communities and International Policy Coordination," *International Organization* 46, no. 1 (1992): 1–35; John W. Meyer, John Boli, George M. Thomas and Francisco O. Ramirez, "World Society and the National-State," A*merican Journal of Sociology* 103, no. 1 (1997): 144–81.

4 Kate Bedford, *Developing Partnerships: Gender, Sexuality, and the Reformed World Bank* (Minneapolis, MN: University of Minnesota Press, 2009), and this volume; Gülay Caglar, *Engendering der Makroökonomie und Handelspolitik. Potenziale transnationaler Wissensnetzwerke* (Wiesbaden, Germany: VS Verlag für Sozialwissenschaften, 2009), and this volume; Stefanie Wöhl, *Mainstreaming Gender? Widersprüche europäischer und nationalstaatlicher Geschlechterpolitik* (Königstein/Taunus, Germany: Ulrike Heimer Verlag, 2007); Carol Lee Bacchi, *Women, Policy and Politics: The Construction of Policy Problems* (London: Sage Publications, 1999).

5 At the time of this writing, the list has grown to 80 manuals and INSTRAW has been merged into UN Women. The URL of the gender training community of practice is http://www.un-instraw.org/gtcop/.

6 The first was produced by a US consulting group, the Futures Group International, together with Religions for Peace, an international coalition of representatives from religions around the world. The second was written by JSI Research and Training Institute, a consulting firm, together with the Reproductive Health Response in Conflict (RHRC) Consortium, which includes a number of NGOs.

7 Nikolas Rose, Pat O'Malley and Mariana Valverde, "Governmentality," *Annual Review of Law and Social Science* 2 (2006): 84–85.

8 International Alert and Women Waging Peace, *Inclusive Security, Sustainable Peace: A Toolkit for Advocacy and Action. Introduction* (Hunt Alternatives Fund and International Alert, 2004a), 5.

9 International Alert and Women Waging Peace, *Inclusive Security, Sustainable Peace: A Toolkit for Advocacy and Action. Conflict Prevention, Resolution and Reconstruction* (Hunt Alternatives Fund and International Alert, 2004b), 25.

10 Ibid., 26.

11 Ibid., 43.

12 Geneva Centre for the Democratic Control of Armed Forces (DCAF), *Gender and Security Sector Reform Training Resource Package. Training Resources on Defence Reform and Gender* (Geneva: DCAF, 2009).

13 Ibid., 4.

14 International Alert and Women Waging Peace, *Inclusive Security, Sustainable Peace: A Toolkit for Advocacy and Action. Conflict Prevention, Resolution and Reconstruction.*
15 UNIFEM, *Advancing Gender Equality Using CEDAW and UN Security Council Resolution 1325: Training Module for Gender Equality Advocates* (New York: UNIFEM, 2003), 154.
16 *Pace* Judith Butler, *Undoing Gender* (New York: Routledge, 2004).
17 UNIFEM, *Getting It Right, Doing It Right: Gender and Disarmament, Demobilization and Reintegration* (New York: UNIFEM, 2004), 21.
18 Catharine A. MacKinnon, *Are Women Human? And Other International Dialogues* (Cambridge, MA: Belknap Press of Harvard University, 2006).
19 Beth Vann, *Training Manual Facilitator's Guide: Multisectoral and Interagency Prevention and Response to Gender-based Violence in Populations Affected by Armed Conflict* (Washington, DC: JSI Research and Training Institute/RHRC Consortium, 2004), 2–8.
20 UNICEF, *Training of Trainers of Gender-Based Violence: Focusing on Sexual Exploitation and Abuse* (n.p.: n.d.), 17, 50ff.

4 Feminist strategies in international organizations

The United Nations context

Carolyn Hannan

- **Goals and strategies**
- **Enabling factors**
- **Challenges**
- **Impacts**
- **Conclusions**

Feminists and their allies in the United Nations (UN) have played critical roles in promoting gender equality at the global level. As an intergovernmental body, the UN provides unique opportunities to actively influence global norm and policy development and promote change at national levels. Working as delegates for their governments, as staff members within the organization, and as representatives of nongovernmental organizations (NGOs), women and men have put new issues on the agenda, highlighted the gender perspectives in existing agenda items, and questioned and lobbied for change in the terms and procedures of negotiation processes. Sharing common values and goals, these three groups of feminists have worked closely together to develop strategies for positive norm and policy outcomes on gender equality and to strengthen monitoring and implementation of these norms/policies at the national level.

Since the UN is an intergovernmental body, insiders can include both feminists working within member state delegations and those working within the organization as bureaucrats—also known as "femocrats."[1] Initially there were very few women bureaucrats and much of the early advocacy and norm and policy work was carried out by delegates together with NGO allies. Gradually as the number of women in the UN grew, femocrats began to play an increasingly important role. Following the first World Conference on Women in 1975, women began to be employed as "women/gender specialists" to focus specifically on gender equality. With the adoption of the Beijing Platform for Action (BPA) and the establishment of the gender mainstreaming strategy in 1995, the number of experts within the UN increased significantly. This section presents some of the strategies developed by feminists working inside the UN bureaucracy over the past few decades, their relevance and impact, as well as enabling factors and persistent challenges to their full implementation.

Goals and strategies

From the founding of the UN in 1945, major goals for feminists working inside the organization have included consistently utilizing the opportunities the organization provides for positive development of global norms and policies and using progress at the global level as leverage for positive change at national levels, by ensuring the incorporation of gender equality as an integral part of all work of the organization. Short-term objectives related to specific issues have changed over time as the UN has developed, knowledge on gender equality around the world has increased, and specific gains have been made on gender equality. In the initial years of the UN, for example, the main objective was to increase knowledge of the situation of women around the world through a focus on research and data collection. As knowledge of the glaring gaps in the human rights of women was gained, the objective shifted to development of specific legislation. A similar shift in immediate objectives can be seen in relation to work on the Convention on the Elimination of all forms of Discrimination against Women (CEDAW).

Immediately after the adoption of the Convention, the key objective was securing ratification by states. As the number of states parties steadily increased, the focus shifted to ensuring full implementation at national levels and timely reporting on progress.

Over the decades, a range of strategies has been developed to influence both intergovernmental processes and the support provided by the UN at national levels. The UN Secretariat is responsible for support to the intergovernmental processes where negotiations lead to the development of global norms/policies and mandates for action; and gender specialists working in the UN Secretariat, for example in the former Division for the Advancement of Women (DAW),[2] have had responsibility to provide support for intergovernmental processes on gender equality—in the Commission on the Status of Women (CSW),[3] the Economic and Social Council (ECOSOC) and the General Assembly—in order to ensure positive development and follow-up of norms/policies on gender equality. Although the work program and resources for the secretariat are determined directly by member states through intergovernmental mandates, there is nevertheless some room for maneuver for femocrats in developing strategies to influence important processes and outcomes. The great majority of gender specialists in the UN work as advisors/catalysts—promoting gender equality in the work of the bodies in which they are placed. In UN funds and agencies providing direct support to states, strategies are developed to promote national-level implementation of global commitments.

The initiative for strategies on gender equality in the UN has come from varying sources—initiatives have been taken directly by states, promoted from inside the UN or proposed by NGOs. The impact of strategies has been clearly enhanced when there is direct collaboration among feminists across institutional boundaries. Strategy development has not been linear and approaches have differed over time depending on the opportunities and needs of particular contexts. This has required considerable flexibility and creativity as well as capacity to adapt to changing situations and new opportunities. Feminists inside the UN have consistently worked to take advantage of opportunities when new issues are taken up or new processes initiated. In the context of the early UN focus on human rights, for example, women pushed for the equal rights of women. With the strong focus on development in the 1960s and 70s and the establishment of the international development strategies, feminists promoted the "women in development" agenda. As part of the increased focus on peace and security issues in the late 1990s, women worked to ensure that the critical issue of women, peace and security was brought to the agenda. Tactics and alliances which worked effectively in one context have not always been useful in others. Shifts in alliances over time can be illustrated in the work on violence against women. Since NGOs were the key actors in promoting attention to violence against women within the UN, femocrats were initially focused on developing alliances with important NGOs in this area. As global interest on violence against women increased, it became strategically important for both femocrats and NGOs to collaborate with member states to ensure the issue became firmly established on the global norm and policy agenda through intergovernmental processes. Feminists working with intergovernmental processes have needed to promptly recognize risks to processes and outcomes. In the context of conservative political environments in the late 1990s, for example, risks for backsliding on established commitments led femocrats to support delegates in consistently utilizing agreed language to reduce risks in negotiation processes.

The role of the DAW in servicing the CSW provided opportunities to influence global norm and policy development. Work in this capacity involved strategic activities such as preparing reports of the Secretary-General, guiding and supporting the work of the bureau of the CSW, organizing substantive elements of intergovernmental meetings and providing advice and support in negotiation processes. These activities provided opportunities to push policy development and promote new issues on the policy agenda. Staff members inside the UN Secretariat worked, for example, to influence the work program of the CSW through submission of menus of possible priority themes, in order to ensure relevant issues became topics for discussion. The development of appropriate working methods in the CSW to ensure continued relevance both internationally and at national levels was another strategic area of work. Following the Fourth World Conference in 1995, the division placed strong focus on strengthening the CSW's role in following up efforts at national levels. This included the organization of interactive expert events as part of the regular annual program of the CSW to raise the level of dialogue and debate and increase exchange among states on national-level experiences. Expert meetings have been an important instrument for bringing diverse actors together to discuss critical issues, gain access to experience in the field across all regions, and provide opportunities for exchange of national-level strategies. Expert meetings are used effectively to bring alternative perspectives—from practitioners and activists from all regions—to the themes addressed by the CSW at its annual sessions. This broadens the debate and facilitates an inclusive approach.

Identifying ways to raise strategically new issues has sometimes been a challenge in intergovernmental processes. Some issues are not immediately recognized as important and considerable time and lobbying may be required before they are taken onto the policy agenda. The importance of attention to men and relations between women and men is a good example of one issue which did not resonate well with states when first proposed as a policy item in the 1990s, but which is now well-established on the intergovernmental agenda. Strategies utilized to ensure that the issue took root included lobbying, networking with experts, exploratory research and the production of a publication, building on work carried out in all regions to increase dialogue with and involvement of men.

Feminists in the UN gave priority to work with the global normative and policy frameworks—CEDAW (1979) and the BPA (1995). Following active involvement in the development of both instruments, feminists working inside the UN Secretariat undertook substantive servicing of the bodies responsible for follow-up of implementation—the CEDAW committee and the CSW.[4] This offered opportunities for influencing roles and methodologies to enhance the impact of the instruments at national levels. Expert panels were introduced in the CSW to provide states with access to globally recognized expertise on critical gender equality issues. The CSW is today an important forum for exchange between states. The fact that states increasingly share information not only on achievements and promising practices but also on persistent gaps and constraints is a positive trend. The regular five-year reviews of implementation of the BPA require states to systematically report on progress made. Similar support by feminists to the continuous development of working methods of the committee of independent experts monitoring the implementation of CEDAW has also had positive impacts on the practical utilization of the Convention at the national level.

To enhance understanding of the relevance and importance of these global frameworks, strategic communication and outreach activities were undertaken by the DAW. Information was disseminated broadly, in accessible forms, using publication, brochures and websites. The recommendations arising from the annual sessions of the CSW are today, for example, produced in an accessible brochure form, in all six languages of the UN, and widely disseminated to promote effective use at national level. DAW organized training programs and convened expert meetings to promote implementation. UN agencies with programs supporting efforts at the national level have developed strategies to support implementation of these global instruments. These include support to participation of diverse stakeholders at meetings of the CSW and the CEDAW committee, dissemination of information on themes and outcomes of the meetings, active use of outcomes in policy dialogue to remind states of their commitments, and efforts to build on norm and policy outcomes in program or project work.

This prioritization of the global frameworks has proven worthwhile. Even 15 years after its adoption, the BPA still provides an excellent global policy blueprint and is regularly followed up in the CSW, ECOSOC and the UN General Assembly. The CEDAW Convention provides unique opportunities to guide and support national-level development through the requirements for four yearly reports to the independent committee of experts, a constructive dialogue between the committee and states parties, possibilities for dialogue with NGOs, and tailored recommendations by the committee for action at the national level. The importance of the instruments and the global follow-up processes are confirmed annually by participants from feminist movements all around the world attending sessions of the CEDAW committee and the CSW. While implementation of both instruments leaves much to be desired in all regions, they have proven indispensable to work on gender equality at the national level and are increasingly recognized by governments as critical instruments.

The shift from perceiving women as a marginalized group needing separate attention to recognition of the importance of addressing their critical roles and contributions in all policy areas was a significant move forward in the UN. This important shift began already in the 1960s in the work on women and development. This led to increasing awareness that women not only need development interventions to improve their own lives but also make significant contributions to development, with major benefits for their families and communities as well as for progress at the national level. Efforts to strengthen women's roles and build on their contributions led eventually to the establishment of the gender mainstreaming strategy which aims to ensure that the priorities, needs and contributions of women as well as men are taken into account in interventions in all policy areas. Feminists working inside the secretariat were instrumental in ensuring adequate understanding of the strategy among delegates prior to the formal recognition of the strategy in the BPA in 1995. Following the adoption of the strategy, collaboration with the ECOSOC in 1997 led to the development of basic principles and guidance on implementation.

Feminists inside the UN bureaucracy have worked to present gender equality issues in a manner that makes them comprehensible and increases understanding among states and senior officials of their strategic importance to the UN vision and goals. Throughout the UN, efforts are made to ensure that gender equality perspectives are systematically addressed in research and data collection, policy and strategy development, and design and implementation of programs or projects in support of state efforts at the national level. A range of strategies has been developed. Almost all UN entities have made gender mainstreaming part of equality policies, developed specific methodologies and tools and carried out training programs.[5] Networking has promoted exchange of information and experiences and facilitated coherence across the system.

Enabling factors

Work on gender equality inside the UN is significantly enabled through collaboration with and support of allies among member states—individual states or groups of states—interested in and supportive of positive change in particular areas. For feminist UN insiders, working closely with individual delegations has been an effective way to increase and strengthen policy attention to critical issues. States are often willing to provide resources for data collection and exploratory research or for expert meetings to bring together expertise from different regions. An illustrative example is collaborative work on violence against women. In the early 2000s, feminists working inside the UN Secretariat were able to capitalize on the strong interest of a number of states, and the commitment of individual delegates in permanent missions in New York. This facilitated work in the General Assembly, leading to an in-depth study of the Secretary-General, a strong resolution with a set of strategic concrete recommendations for action at national level, and strategic follow-up work within the UN, including the development of a global database providing access to information on work around the world to eradicate violence. Similarly in the late 1990s, close contact between the secretariat and delegates committed to the gender mainstreaming strategy led to the establishment of a cross-regional group of states—the "Friends of Gender Mainstreaming." The group worked effectively to promote and support gender mainstreaming in the work of the UN. The "Friends of" concept has since been used in other contexts, for example on women, peace and security and violence against women.

The importance of committed member states willing to defend critical concepts and issues can also be illustrated by what happens when such support is removed. While feminists working inside the UN could previously rely on the unwavering stance of the European Union (EU) on sexual and reproductive rights and health in negotiation processes, the lack of an EU position on this issue since the EU enlargement process means increased vulnerability around the issue. EU member states are no longer able to negotiate on this issue as a bloc and this has opened up the possibility for individual states to put forward positions that are not in line with earlier strong support for the issue coming from the EU as a whole.

A further enabling element is the collaborative efforts of feminists across boundaries—that is, those feminists working in member state delegations, the UN and NGOs. Effective collaboration provides enhanced potential to create momentum for achieving common goals. This can be illustrated in the area of women, peace and security. In the 1990s, the changing nature of conflicts, with increased attacks on civilians and use of violence against women as a weapon of war, forced a new approach to peace and security issues within the UN. Women in different contexts took advantage of this opportunity. Using a comprehensive range of strategies, including lobbying of member states and senior UN officials and provision of critical information, they ensured that the Security Council began to address issues of gender equality. This led to the path-breaking Security Council Resolution 1325 (SC 1325) in 2000. Seminars, conferences, and expert meetings have been used strategically to promote focus on the implementation of the resolution and monitoring of outcomes.

The existence of strong active feminist NGO movements at different levels is a critical element. The inputs of NGOs have enhanced the work of internal advocates—energizing UN debates in critical areas and increasing the visibility of gender equality. A great deal of the sustained attention and achievements in a number of critical areas has been due to their efforts. Their role in advocating and monitoring national-level implementation of the commitment made by states in the global arena is particularly important. NGO networks working at international, regional and local levels were instrumental in securing attention to violence against women in the UN and making clear the crucial links between violence and human rights. Violence against women was not on the global agenda in the 1970s and was therefore not included in the articles of CEDAW (1979). Attention to violence in the Third World Conference on Women in Nairobi in 1985, initiated by women's groups and networks, created momentum that led to the adoption of the Declaration on the Elimination of Violence Against Women in the General Assembly in 1993. Violence against women was then established as one of 12 critical areas of concern in the BPA in 1995 and states began to be more effectively involved in this area. These NGO efforts supported the work of internal advocates.

Feminist NGOs have played an equally important role in building on UN global level achievements at regional and national levels. In the area of women, peace and security, for example, NGOs played a key role in dissemination and use of SC 1325 at national levels, including through translating the resolution into more than 70 languages, demanding accountability of states and pushing for development of national implementation plans. NGOs took strong initiatives, with support of feminists in the UN, to use the resolution to create momentum at the local level in a manner which would not have been possible for the UN. As a result, the Security Council has significantly increased interaction with women's peace and security networks at global, national and local levels over the past decade.

Effective work within international organizations such as the UN is also enabled by support from leaders at top and middle levels. This support needs to be sustained, highly visible and clearly moving beyond political rhetoric. In particular, the active and visible leadership of the Secretary-General on gender equality issues is critical, through, for example, initiatives such as the current Secretary-General's global campaign on violence against women (Unite to End Violence against Women, 2008–15). Unfortunately there is much to be desired in terms of senior-level support for gender equality in the UN. Some notable exceptions include the International Labour Organization (ILO) where explicit expressions of commitment by the director-general strengthened the work of internal advocates, leading, for example, to innovative work with gender audits.[6] In the Department of Peacekeeping Operations (DPKO), statements by an under-secretary-general to the Security Council in 2005–7 clearly identified failings in the department and the required changes to ensure implementation of SC 1325 on women, peace and security.[7] The fact that senior managers make strong policy statements on gender equality does not, however, ensure that there is consistent strong and proactive leadership within organizations. Even when top-level endorsement of gender equality is in place, commitments can be undermined by lack of engagement of middle-level management. Stronger accountability mechanisms are required to develop supportive leadership at all levels.

Challenges

In developing and implementing strategies, feminists working inside the UN have faced numerous serious challenges. These include, among other things, challenges in relation to concept development; the complexity of global standard setting in the context of conservative political climates and differing levels of commitment among states; weak follow-up by states on agreements achieved; and serious gaps in prioritization, resource allocation and leadership on gender equality. Addressing these challenges requires creativity, flexibility and strong alliances.

Since the inception of the UN, concepts have presented one such challenge, particularly in relation to moving the normative and policy agenda forward in intergovernmental processes. The struggle with concepts is not just a game of semantics as sometimes presumed but can have significant consequences for the direction of norm and policy development. The importance of conceptual clarity was, for example, clearly recognized by the first feminists working with the UN in member state delegations and in NGOs. They worked strategically to ensure that concepts used in the founding Charter of the UN, outlining the overall organizational vision and goals, were conducive to the promotion of gender equality. They carefully guarded the gains made in the Charter in the development of the Universal Declaration of Human Rights. This required considerable advocacy and lobbying.[8] This strategic work was continued with support of feminists working inside the organization.

The complexity of concepts creates challenges. In promoting clarity of understanding to facilitate both norm and policy development and implementation at national levels, the challenge has been to utilize concepts with universal validity which at the same time respect historical and cultural contexts. Promoting broader definitions of violence against women, for example—finding a means to frame violence as a human rights violation while encompassing the many different forms of violence around the globe and unifying the wide range of activist groups working on specific practices—has been of critical importance. Concepts are not static but are continuously evolving and subject to debate, contestation and reinterpretation. Conservative groups can utilize demands for conceptual clarity in non-constructive ways, for example as a means to derail discussions and water-down approaches. Prior to the Fourth World Conference in 1995, for example, efforts were made by conservative groups to replace the word "equality" with "equity," on the grounds that "equity" implied fairness and that what was considered fair treatment for women differed between cultures and countries. A strong stance by feminists in the CSW resulted in the retention of the stronger concept of "equality." Similarly, a non-constructive debate on the concept of "gender" was initiated by an alliance of conservative forces, including the Vatican, in the CSW prior to the Fourth World Conference, to hinder progress in the development of the BPA. Again, the CSW took a strong stance in support of "gender" as a useful concept in the work of the UN. The non-constructive debate on "gender" has, however, been a recurring problem. As recently as March 2010, conservative forces in the CSW again mobilized to create a false conflict around the use of "gender."

In the conservative political climate after 1995, serious challenges to providing effective support to intergovernmental processes emerged. It became difficult to maintain positions from the Fourth World Conference, let alone move forward in policy development. The growing strength of anti-abortion and conservative "pro-family" alliances—including conservative NGOs—with clear goals of back-tracking on established policy commitments and derailing negotiation processes[9] resulted in the important feminist strategy to keep to "agreed language" rather than attempt to negotiate new policy positions. As a result of this approach, policy positions were successfully maintained in the five-year reviews of the Programme of Action from the International Conference on Population and Development (1999) and of the BPA from the Fourth World Conference on Women (2000).[10] However, because of clear risks to maintaining existing policy positions experienced in these review processes, committed states made the difficult decision not to agree to negotiation of new policy outcomes in the 10- and 15-year reviews in 2005 and 2010. The persistent conservative opposition has meant therefore that significant policy development has been constrained in the UN for more than a decade. Feminists in the UN, many states and NGOs have opposed the idea of a fifth world conference; a conference outcome less than that obtained in Beijing in 1995 must be avoided.

A clear challenge in the work on gender equality in the UN is the fact that there is not complete agreement among states on all issues, despite the unanimous adoption of the BPA and the ratification of CEDAW by almost all states. This increases the complexity of global standard-setting. Differences in the situation and position of women in states can lead to debates and controversies, even among feminists, over the most urgent priorities and most appropriate strategies. There can also be considerable resistance by individual states or groups of states on particular issues for political reasons. Controversial issues which have consistently met with opposition in negotiation processes include sexual and reproductive rights and health, abortion, sexuality and the definition of "family."

Tensions can emerge between claims of national sovereignty and the universality of global norms and policies which seriously constrain negotiation processes. Insidious trends to water-down commitments have included attempts to insert reference to the need to take traditional customs/culture into account as a means to limit the applicability of recommendations, or reference to the need for global policies to be implemented in the context of national strategic development plans (where this implies that national values and goals should take precedence over global norms and policies). Gender equality can also be held hostage to other political issues. In 2008–9, in the context of the ongoing coherence reform process in the UN, the finalization of the reform of the gender equality work of the UN was significantly delayed because some states would not agree to move forward until other areas of the reform package were addressed.[11]

Weak follow-up by states to agreements and commitments made in intergovernmental processes creates challenges. Failure to implement commitments at national levels is a serious problem requiring innovative strategies in UN entities with field presence and national- or local-level NGOs to increase accountability. States may also fail to systematically follow-up and react when important recommendations directed at the UN are not adhered to. An illustrative example from the Security Council refers to recommendations that reports of the Secretary-General to the council should include information on gender equality. When these recommendations are not followed, states unquestioningly accept the reports submitted, which has the effect of reducing the recommendations to mere rhetoric.

Although feminists working inside the UN have made consistent efforts to take advantage of new opportunities, for example when new issues are taken up or new processes initiated, these efforts have often been constrained by failings in planning processes to adequately consider gender equality and build on commitments and achievements made. Challenges, for example, emerged in finding ways to positively utilize the Millennium Development Goals (MDGs). While the Millennium Declaration adopted by heads of state at the Millennium Summit in 2000 was in many ways a positive document from a gender equality perspective and created significant global momentum, the MDGs subsequently developed by the UN were extremely weak. The eight goals were a subset of measurable goals established in the global conferences in the 1990s, mainly focused on traditional basic needs, with narrow targets and indicators. Although a specific goal on gender equality and empowerment of women (MDG3) was established, as well as a goal on maternal mortality (MDG5), feminists around the world were outraged that the priorities and focus of the BPA were largely ignored, in particular human rights, violence, and sexual and reproductive rights and health. In many ways, the adoption of the MDGs was viewed as a major setback for gender equality. As the new framework began effectively to mobilize governments, international organizations and NGOs around the world, feminists struggled to find ways to turn the MDGs into a positive force for gender equality. There has been increased recognition that the focus on implementation and on reporting does provide important opportunities for increasing the visibility of gender equality in national development planning processes, and that the MDG framework enhances monitoring and reporting, which facilitates the development of new alliances and improves access to resources.[12]

The recent reform process on gender equality identified serious challenges related to the prioritization of and allocation of dedicated resources for gender equality work in the UN. The fragmentation of the issue between small, seriously under-resourced entities was compounded by the lack of consistent, strong and pro-active senior-level leadership. Recommendations emerging from the reform included the merging of existing specialist entities into one entity—the UN Entity for Gender Equality and the Empowerment of Women (UN Women), the upgrading of the position of head of entity to under-secretary-general level, and the promise of significantly increased resources.

The glaring failure of the UN to achieve Article 8 in the UN Charter which requires all positions in the UN to be equally open to women and men also constitutes a challenge to effective work. This is not because of an expectation that all women employed would be feminists, but because the under-representation of women in the UN contributes to the creation of a male-dominated environment with values, norms, attitudes and procedures which are not conducive to promotion of gender equality. Serious attention to gender balance in the secretariat was first noted in the early 1970s with the appointment of a woman assistant secretary-general in 1972. Since then, the number of women has increased but the goal of equal representation has not been achieved. Progress in increasing the number of women in high-level decision-making positions has been particularly slow. According to recent statistics, for example, women make up only 23 percent of the highest positions in the UN—under-secretary-generals.[13] In this context, it is interesting to note that while the number of women in member state delegations has increased over the decades, the number of women as permanent representatives of their countries has remained extraordinarily low, never reaching more than 20 of 192 states. This is probably related to the fact that ministries of foreign affairs have notoriously poor records on gender balance.[14] There may be a link to the low representation of women in senior management levels in the UN. Member states which do not promote women in decision-making positions in their own bureaucracies cannot be expected to give this high priority within the UN.

Some challenges can be identified in relation to NGO roles in the intergovernmental context. Although the UN is a member state-based organization, opportunities have always been provided for NGO participation, including in intergovernmental processes. Formal NGO participation in the CSW needs, however, to be significantly broadened. While there are strong advocates for enhanced NGO participation, such participation is not without contestation. A significant number of states actively oppose increased NGO involvement. In addition, the existing participation of NGOs could be significantly strengthened. While large numbers of NGO representatives attend the annual sessions of the CSW, experience has shown that not all have capacity to participate fully and many do not take back relevant information to the national level, which limits the impact of their participation. Many NGOs are narrowly focused on one or two issues and do not always have the broader gender equality context in focus. Much more could be done by New York-based NGOs, which have year-round access to and knowledge of the UN and its processes, to strengthen understanding among NGOs at regional and national levels of UN processes, their importance and ways outcomes can be used at national level. This would enhance the impact of the work inside the UN. A further critical challenge for the UN concerns ways to engage women's groups and networks at national and local levels which will never attend sessions of the CSW in New York but could benefit from access to information and participation in a broad global network. Such groups and networks make significant inputs towards ensuring effective implementation of global norms and policies at national and local levels which enhances the work carried out in the UN.

Impacts

Despite major challenges, there have been many significant achievements. In assessing impacts of feminist strategies it is difficult to identify successes since much of the work on gender equality is on-going and evolving. In all areas where success can be recorded, much more remains to be done. It is also difficult to attribute causality for gains made since all work on gender equality has involved collaboration between member states, UN staff and NGOs. There has been significant institutional learning and strategies utilized have brought about many changes in the UN—in values, norms, structures, procedures and outcomes. Change can also be seen in increased levels of knowledge, awareness and commitment among states and UN officials. Most success has been achieved in the development of critical norms and policies. Remarkable achievements include the development and adoption of the BPA and the CEDAW Convention and its Optional Protocol which adds an individual complaint procedure as well as an inquiry procedure to the Convention's monitoring mechanism. that enables women to draw the Committee's attention to concrete cases of violations of women's rights as laid out in the Convention. Since the Fourth World Conference in 1995, the CSW has also made some policy advances, despite the conservative political climate, on a number of critical issues, such as the role of men and boys, financing for gender equality, women's political participation, the situation of girls, and climate change and natural disasters. In addition, important policy developments have been made on human rights and violence against women in the context of the General Assembly and on women, peace and security in the Security Council. The Millennium Declaration and the outcome of the 2005 Millennium Summit also provided important policy guidance on gender equality. There remains, however, a huge gap between these global instruments and implementation at national levels, which remains one of the major challenges facing the UN today, creating risks in relation to authority and relevance.

Notable impacts of feminist strategies include the increased relevance of the CSW as a result of continuous development of working methods and strong focus on national-level implementation—which is evidenced by high levels of participation and ministerial-level representation at annual meetings. The five-year reviews of implementation of the BPA (2000, 2005 and 2010) have also been successful, high-profile events, despite the less than optimal political climate. This indicates the continued importance attached to the BPA as the global policy framework, and is evidence of the critical role of the UN in bringing together diverse stakeholders. It must also be considered a major achievement that it has been possible to maintain the policy positions in the BPA and strengthen the impact of CEDAW over the past decade, despite significant attacks from conservative forces and concerted efforts not only to hinder new development but also to backtrack on existing commitments.

While initially work on gender equality was carried out solely by the CSW, as a result of important catalytic work most intergovernmental bodies in the UN today pay attention to gender equality, including ECOSOC, the General Assembly and the Security Council. Similarly, it is no longer only gender specialists who are expected to work with gender equality—the gender mainstreaming strategy requires that all personnel take gender equality into consideration throughout their work. While gaps still exist in relation to both intergovernmental processes and the capacity and commitment of staff to work effectively with gender equality, a significant shift of responsibility can be noted.

Many serious gaps in the impact of feminist strategies on the UN, including on structures, procedures and outcomes, however, exist. The efforts of feminists have, for example, had limited impact on the macroeconomic environment. Important policies and strategies in critical areas recently on the UN agenda, such as climate change and the financial crisis, have been discussed and negotiated with almost no consideration of gender equality implications. Among major failures can be noted the fact that, despite considerable efforts across the entire UN system, full implementation of gender mainstreaming has never been achieved—not in the UN itself or in any state. In no context have the required political support and resources been invested to ensure the transformative potential of the strategy. A serious constraint to full implementation is the lack of visible leadership from senior and middle-level management, as well as resulting failures to establish effective accountability mechanisms. Despite training programs, capacity is also underdeveloped in many UN organizations. Misperceptions continue to hinder effective implementation. While the value of gender mainstreaming has been increasingly questioned, no viable alternative has ever been presented. It is premature to declare the failure of mainstreaming but high time for serious discussion of ways to accelerate systematic and effective implementation.[15]

One of the most persistent problems is the continued separateness of gender equality efforts. While most organizations have gender equality policies, not enough has been done to ensure gender equality is systematically incorporated as a goal into other critical organizational policies and strategies, for example, on poverty reduction and environmental sustainability as well as on specific sectors, such as energy and water. This creates a serious gap in organizational mandates and results in continued failure to work with gender equality as an integral part of all issues and sectors. The problem of separateness can also be illustrated by inadequacies in methodology development. Although useful mainstreaming methodologies and tools have been developed, gender perspectives are not necessarily incorporated into sector methodologies and tools. While gender analysis may be undertaken as a separate analysis, gender perspectives are not systematically considered in other relevant analyses—for example on poverty and environmental impact.[16] Coordination across the system in the development of strategies and methodologies has been lacking and this has created unnecessary complexity in work at national levels. The mainstreaming strategy requires a shift of responsibility from specialists to all personnel, especially management levels. The role of gender specialists should be to support managers to fulfill their management and leadership responsibilities for promoting gender equality. To function effectively in this critical role, specialists need clear mandates, strategic location, direct access to senior management and adequate resources. It would be timely for feminists to reconsider the appropriateness and relevance of strategies for working inside the UN and assess the effectiveness of efforts made.

Conclusions

While noting that the UN remains a male-oriented environment, that UN decision-making processes can be slow and that decisions may lack implementation, Indian feminists recently pointed out the concrete gains from feminist interventions in the organization: "… the United Nations has played a unique role: as a catalyst for change, as a global standard-setter for the eradication of gender discrimination; as a forum for debate; and as an unparalleled source of balanced, comprehensive data on the status of women worldwide."[17]

Nonetheless, more than 60 years of UN efforts have not yet produced the desired outcomes on gender equality. Despite increasingly positive rhetoric, the promotion of equality is not perceived as central to all areas of work on human rights, development and security—the three pillars of the UN. The actions required for moving the gender equality agenda forward are often very well known—the challenge is ensuring effective implementation.[18]

The new era signaled by the establishment of UN Women should involve much more than a consolidation of former entities. UN Women must address the challenges outlined and build on the progress already made by the UN. The entity is charged with providing global vision, new strategic direction and creative leadership on gender equality. As a result of the long drawn-out reform process, expectations of what can be achieved are huge—perhaps unreasonable. UN Women will need significant political support, as well as enhanced alliances with feminists in member state delegations and women's groups and networks, including in academia, if it is to be able to fulfill its role and ensure that the UN can inspire and lead the world on gender equality and empowerment of women.

Notes

1 The term femocrats is used in this section to refer to feminists working inside the UN organization as bureaucrats. This term was first introduced in Australia in the 1970s to describe feminists who entered women's policy positions in government. See: Marion Sawer, *Femocrats and Ecocrats: Women's Policy Machinery in Australia, Canada and New Zealand* (Geneva: UNRISD, 1996); and Hester Eisenstein, *Inside Agitators: Australian Femocrats and the State* (Philadelphia, PA: Temple University Press, 1996).

2 As of 1 January 2011, the four previous entities in the UN which focused exclusively on gender equality—the DAW, UNIFEM, INSTRAW and the Office of the Special Advisor to the Secretary-General—have been dissolved as part of an effort to strengthen the work of the UN on gender equality. Their mandates have been incorporated into a new entity, UN Women. The work of DAW on norm and policy development in intergovernmental processes continues in UN Women.

3 The CSW is the intergovernmental body mandated to support and follow up the implementation of global norms and policies on gender equality, including in particular the BPA. It meets annually and brings together the expert bodies at national level responsible for the promotion of gender equality.

4 The Committee on the Elimination of Discrimination Against Women is a group of 23 elected independent gender equality experts, which is responsible for the monitoring of implementation of the Convention on the Elimination of All Forms of Discrimination against Women (CEDAW). The committee reviews official implementation reports and holds constructive dialogue with representatives of the state party. It also meets with NGOs and UN entities to gain access to relevant information on implementation at national level, before providing concrete recommendations to the state party on the actions required to improve implementation of the Convention.

5 Carolyn Hannan, "United Nations Gender Mainstreaming Strategy: Achievements and Challenges," *Gender Studies and Policy Review* 1 (2008).

6 Juan Somavia, director-general of ILO from March 1999 to the present.
7 Jean-Marie Guehenno, former under-secretary-general of DPKO in the 1990s.
8 Devaki Jain, *Women, Development and the United Nations: A Sixty Year Quest for Justice* (Bloomington, IN: Indiana University Press, 2005), 12–20.
9 Gita Sen, *Neolibs, Neocons and Gender Justice: Lessons from Global Negotiations* (Geneva: United Nations Research Institute for Social Development, 2005), 8–10; and Margaret Keck and Kathryn Sikkink, *Activists Beyond Borders: Advocacy Networks in International Politics* (Ithaca, NY: Cornell University Press, 1998), 165–98.
10 Sen, *Neolibs, Neocons and Gender Justice*, 8–10.
11 The coherence reform process was initiated by Secretary-General Kofi Annan following the 2005 Millennium Summit. It included five areas where increased coherence was promoted—business practices, UN delivery mechanisms at country level ("Delivering as one"), finances, governance, and gender equality. In 2008–9, three of the elements in this coherence reform package—finances, governance and gender equality—remained unresolved. Although considerable work had been done on gender equality and many member states were ready for conclusion of the process, other states refused to finalize an outcome on gender equality until further work was done on governance and finances.
12 Carolyn Hannan, "Empowering Women: Ten Years After the Beijing Conference," *Georgetown Journal of International Affairs* (Summer/Fall 2006): 173–82.
13 *Composition of the Secretariat: Staff Demographics. Report of the Secretary-General* (General Assembly document A/65/350), 8 September 2010.
14 Reliable, comparable global statistics on gender balance in ministries of foreign affairs are not available.
15 Carolyn Hannan, "Gender Mainstreaming: Failings in Implementation," *Kvinder & Køn Forskning* 1 (2011): 57–60.
16 Hannan, "Gender Mainstreaming," 59.
17 Devaki Jain and Shubha Chacko, "Unfolding Women's Engagement with Development and the UN: Pointers for the Future," *Forum for Development Studies* 1 (2008): 10–18.
18 Hannan, "Empowering Women," 173–82.

5 A perspective on feminist international organizing from the bottom up

The case of IGTN and the WTO

Mariama Williams

The ability of gender advocates to lobby effectively the World Trade Organization (WTO) and other institutions of global governance depends on the strategies they pursue to build awareness around gender issues, the creation of sharply focused policy angles and on well-glued alliances and strong collaborative partnerships. In order to increase their visibility and lessen their disadvantage of being newcomers to the areas of macroeconomics, finance and trade, feminists focused on forging strong partnerships and cross alliances among themselves and with the broader civil society movement. In 1999, over 45 gender advocates and women's organizations launched the International Gender and Trade Network (IGTN), which arguably in subsequent years became the most influential gender advocacy organization in the international trade policy arena.

This chapter focuses on the dynamics of gender organizing in the trade area through an examination of the work of IGTN from 2000 to 2010. The aim is to explore what lessons this holds for enhanced work on gender equality and women's economic empowerment.

Women organizing on trade issues: the emergence of IGTN

The Uruguay Agreement, which gave birth to the WTO, concluded in 1994, and the WTO became operational in 1995, the same year states adopted the Beijing Platform for Action. The Uruguay Agreement greatly expanded the scope of multilateral trade governance to include new areas such as agriculture and intellectual property rights. It also expanded the scope for enforcing multilateral trade discipline, affecting areas that hitherto were strictly under the domestic management of governments. The new trade liberalization agenda also negatively impacted the budgets of developing countries, especially those with a high dependence on cross-border taxes for revenue.

The new dimensions of trade agreements and the widened scope of WTO operations have tremendously adverse impacts for women's control of domestic markets in many countries, their livelihood and lives. Thus the WTO immediately became the focus of attention of a small subset of the women's organizations that had been part of the economic justice caucus during the Beijing preparatory process. They participated in the Women's Caucus of the WTO Meeting of Trade Ministers Meeting (also called a Ministerial Conference) in Singapore in 1996 and constituted themselves as the Informal Working Group on Gender and Trade (IWGGT).[1] Initially, the IWGGT focused on understanding the dynamics of the new entity and sought to find windows for integrating gender concerns. For example, members of the IWGGT met informally with the Director of the WTO's Trade Policy Review Division (TPRD) and his staff in March 1997 to discuss the possibility of incorporating gender-sensitive perspectives in trade policy reviews. The TRPD was very receptive and a great deal of time was devoted to exploring in detail the relationship between gender and trade. But it was clear that TPRD was inadequately resourced to deal with the issues and felt that it would be too much of a stretch of its mandate. There was, however, an invitation for future sharing and exchange of information on this issue.

The IWGGT was led by Women in Development Europe (WIDE), given its location in Brussels and its proximity to Geneva. WIDE also was the only women's group with a specific working group on alternative economics and a designated program officer working on alternative economics and trade policies. WIDE's coordination was supported by Women Working Worldwide (WWW) and the women's focal point of the International Confederation of Free Trade Unions in Brussels (now renamed the International Trade Union Confederation) among others. WIDE also collaborated with the Southern feminist network Development Alternatives with Women for a New Era (DAWN).

In 1998, the Global Woman's Program of the Center of Concern (Washington, DC), under the leadership of Maria Riley in collaboration with Peggy Antrobus (DAWN Caribbean), created the Women's Strategic Planning Seminar on gender and trade which focused on building understanding around trade issues for a wide variety of women's groups through online discussions. This seminar and a worldwide civil society coordination process, geared at halting rampant trade liberalization, helped to trigger the formation of a Seattle women's list and steering committee to more effectively coordinate women's nongovernmental organization (NGO) participation at the WTO Seattle meeting. As a result, women's groups had significant presence and visibility at the WTO Seattle Ministerial Conference (1999). This ministerial conference was the third and most contentious meeting since the WTO came into force. Due to the strong resistance of developing countries—which were by then beginning to experience the negative effects of trade liberalization—and well-coordinated protests by global civil society as part of the global campaign, "no new round, turn around,"[2] the meeting collapsed spectacularly. These moments were crucial building blocks for strengthening women's ability to influence trade policy at the multilateral, national and local levels. In the organizing towards the Seattle meeting gender advocates were key participants in the numerous global teleconferences, strategy sessions and teach-ins on trade issues in their national and regional context.

After the Seattle WTO debacle, the Women's Strategic Planning Seminar meeting in Grenada brought together over 48 gender activists and women's organizations working on economic and trade issues in December 1999. A number of participating organizations, such as the Center of Concern, DAWN, WIDE and the Caribbean Association for Feminist Research and Action (CAFRA), agreed to create the International Gender and Trade Network as an independent and autonomous advocacy network. The aim was to promote gender analysis of trade policy and trade agreements and to build knowledge about the linkages between multilateral, regional and bilateral trade agreements and their impacts on women's employment, the informal economy, and social reproduction.

From very early on IGTN focused on the need to work towards alternative models to the current neo-liberal market-driven framework particularly at the national and regional levels. The Steering Committee of IGTN considered it absolutely critical that a feminist gender analysis grounded in women's rights and the broader human rights framework—particularly economic, social and cultural rights—was maintained as central to the network's triple pillars of research, advocacy and trade literacy. IGTN operated from the perspective that the trade agenda was increasingly venturing into new spheres of rule-making that brought it into direct contestation with other areas of global governance such as human rights and the environment. It was also coming into conflict with domestic, behind-the-border issues such as competition policy, government procurement and domestic regulation with impact for the provisioning and access to necessities such as water and healthcare services. IGTN argued for a much more narrow trade focus of the WTO, including the lessening of trade disciplines on agriculture, especially as it related to food. Accordingly, from the beginning IGTN focused on subjects such as food security (later articulated in terms of food sovereignty),[3] access to basic social services and the importance of governments having a high degree of control over their domestic regulation[4] and procurement policies. This control enables governments to undertake remedial measures such as affirmative action programs and to support the household and reproductive sectors if deemed necessary. It is these very topics that are now the bone of contention and important elements underlying the slowly dying Doha Round of trade negotiations.

DAWN—with almost 20 years of policy-oriented research and global advocacy and activism in the areas of political economy of globalization, sexual and reproductive rights, sustainable livelihoods, environmental justice, political restructuring and social transformation—and the Women's Project of the Center of Concern—a United States-based organization centered in catholic social justice and feminist ethos and committed to solidarity with the peoples of the developing world, especially poor women—were pivotal in lending both their names and human resources to the creation of IGTN. They helped brand IGTN as a feminist and Southern-led entity and ensured that it was accepted as an important organization working in tandem with other veteran women's organizations. The Woman's Program of the Center of Concern was the secretariat for IGTN for its first six years and helped to foster the growth and development of the network. The Ford Foundation (New York office) provided the core institutional and programmatic funding for the network for its first six years. Ford's regional offices and other funding agencies supported the development of regional nodes in Africa, Asia and Latin America.

After 10 years of lobbying on gender and trade, international institutions increasingly are looking at the issue, but gender and trade activism at an international level has weakened due to declining funding to support global activities of women's NGOs, especially in the area of macroeconomics and trade. Like many other NGOs, IGTN has had to close its office in Geneva. The continuing moribund status of the Doha Round of negotiations, now in process since 2001, is making trade work less attractive to funding agencies. IGTN is currently in a state of dormancy due to lack of funding.

IGTN and women organizing on trade issues: strategies

IGTN connects feminist gender specialists, groups and networks of feminists, and activists focused on economic rights and social justice working in tandem with economists, in countries around the world. It has conducted research, advocacy and popular literacy training on trade and economics in Africa, Asia, Caribbean, Central Asia, Europe, Latin America, the Middle East and Gulf, and North America. IGTN members provide technical information on gender and trade issues to women's groups, NGOs, social movements, governments, and academic institutions. IGTN also acts as a political catalyst to enlarge the space for a critical feminist perspective and global action on gender, trade, and globalization issues.

The networking strategy of IGTN

IGTN is a transnational formation of gender advocates with a global agenda and regional and national threads. It blends the priorities and concerns of feminist organizations working at the global level into the mandate of a singularly focused entity. These organizations had developed a high level of comfort with and awareness of each other's focus on economic justice, through their joint collaboration during the preparatory processes of the Fourth World Conference on Women in Beijing and the World Summit on Social Development in Copenhagen. Each organization retained its autonomy and identity but lent its human resources and reputational capital to the creation of IGTN as an independent and autonomous entity. IGTN had its own governing structure, leadership, decision-making and funding processes. Critical to the success of IGTN was that it drew strength from the political, social and academic work conducted in other spheres and by members of the partner organizations. It was remarkably focused on elaborating the gender dimensions of trade analysis, moving in tandem with multilateral, regional, and bilateral trade agendas. Regional chapters were free to follow national and regional agendas as they saw fit.

IGTN researchers are very much integrated into academic networks such as the International Working Group on Gender, Macroeconomics and International Economics (GEM-IWG) (see Gülay Caglar in this volume). Researchers are also integrated into their own networks such as DAWN and WIDE, which constantly reinforces a linking to other gender issues. IGTN's research is specifically focused on influencing macroeconomic policies. It aims to provide continuity to existing research processes about issues of gender and trade; to exchange existing research and work methodologies among focal points and among IGTN regions; and to contribute to the development of researchers' networks in the regions. Its capacity building on trade and economic literacy also adds value to existing women's organizations and contributes to the growth of new organizations in some regions. It disseminates information on trade agreements amongst the women's movement, develops programs and mechanisms to increase capacity building, and produces updated training materials and tools.

IGTN also engages in advocacy and political articulation with trade officials and international NGOs working on trade and economic development. At the international level, this aspect of the work was led by its global secretariat and its Geneva-based office, which coordinated work with trade negotiators and directly interfaced with trade organizations such as the WTO and the United Nations Conference on Trade and Development (UNCTAD). Over time, IGTN became an accepted player among the trade and development NGO community and thus gained visibility in the WTO arena. IGTN took part in many networking meetings of NGOs, which increasingly had a physical presence in Geneva. For example, IGTN from very early on closely collaborated with pre-eminent development NGOs such as the Third World Network and Oxfam in the struggle to the incorporation of new issues such as competition policy, government procurement policy and investment into multilateral trade negotiations.

A key focus of IGTN's networking strategy was to promote regional chapters in Asia, Africa, Latin America, Caribbean and North America as well as to work with new regions (Central Asia and the Middle East) requesting to be part of IGTN. Apart from deepening work at the regional level and elaborating regional-specific impacts of WTO-led trade liberalization, regional delegations came to Geneva when issues of particular concern for a region were high on the WTO negotiations agenda. Such delegations were able to deepen their political contact with representatives of regional and sectoral coalitions of countries[5] as well as meet with negotiators from their own countries, which in many cases was not possible at home. Moreover, IGTN closely collaborated with women's movements, gender advocates and networks like the Hemispheric Social Alliance, which was contesting regional trade agreements such as the proposed Free Trade Areas of the Americas (FTAA) and the North American Free Trade Area (NAFTA). IGTN regional chapters also worked on various bilateral agreements of the European Union (EU) with countries such as India and Mexico.

IGTN, through its Geneva office, provided timely and easily understandable updated information on the pace and turns of the WTO negotiations from a gender perspective. This was accomplished through the bi-monthly IGTN electronic Bulletin, occasional updates from Geneva, working papers on specific topics such as on financing for development, aid effectiveness and aid for trade (at least three times a year), and special reports on different aspects of the trade negotiations agenda.[6] IGTN also participated in social movement activities such as the World Social Forum, EU summits, Latin American summits, and UNCTAD conferences. With the outbreak of the financial crisis, IGTN had a strong presence, along with the Association for Women's Rights and Development (AWID), DAWN and WIDE, in fostering strong women's and gender activists' participation in civil society strategizing around G-20 meetings, and UN-related activities on the financial crisis.

Since its establishment, IGTN has networked with women's groups focused on UN-centered activism, in addition to gender focal points inside the UN system. Institutions such as the United Nations Development Fund for Women (UNIFEM, now part of UN Women) and UNCTAD were critical first movers in focusing official attention on the rising importance of women to the new trade agenda. UNIFEM led the way, having implemented a number of research projects on the subject in South East Asia.[7] UNCTAD published one of the first compendium volumes of gender and trade case studies[8] as the manager of the UN Inter-Agency Network on Gender Equality's (UN-IANGE) Task Force on Gender and Trade. However, the work of institutional players tended to be one-off activities with not much systematic and sustained follow-up. It also, for the most part, did not involve any significant outreach to either policy makers or support to women's organizing efforts. For example, not until 2009 did the UNCTAD office begin to work seriously on mainstreaming gender into trade policy as part of its presentation to the UNCTAD Trade and Development Board. UNIFEM was not a visible presence in any fora focused on trade policy and trade negotiations.

While in the 1990s and early 2000s gender and trade activism was met with a great deal of skepticism by trade negotiators and development NGOs, today governmental actors like ministries of trade in India and Uganda are undertaking their own research on gender and trade. The Indian Ministry of Commerce with support from the Commonwealth Secretariat has commissioned studies on the gendered impacts of tariff reduction and Uganda has gender sensitized its national export promotion strategy.[9] Additionally, important institutional players such as UNCTAD, the Food and Agriculture Organization (FAO), the United Nations Development Programme (UNDP), bilateral developmental assistance agencies like the Canadian International Development Agency (CIDA), and intergovernmental forums such as the Commonwealth Secretariat are all strong supporters of gender issues in trade policies.

IGTN methodology: critical perspectives on gender mainstreaming

Even as IGTN was focused on trade, it remained wedded to the concerns, priorities and activities of the broader women's movement and focused on realizing the objectives codified in the Beijing platform and the Convention on the Elimination of All Forms of Discrimination Against Women (CEDAW). But given its advocacy against the expansion of the WTO into broader areas, IGTN was quite wary of pushing a gender mainstreaming agenda inside the institutional orbit of the WTO. Focusing its efforts on mainstreaming gender into trade policy, trade agreements and trade negotiations, IGTN resisted pressures to advocate for incorporating a gender mechanism into the WTO. For IGTN, mainstreaming gender into trade policy means explicitly to make links between trade policy and women's empowerment and gender equality and advocating for provisions in trade agreement and trade negotiations that work towards these objectives and, thus, are supportive to women in their multiple roles in trade. This came to a head at the Cancun Ministerial (2003) when other gender advocates sought to push for the inclusion of a gender focal point and gender mechanism inside the WTO. IGTN pushed back strongly; it was not interested in engendering the WTO as it was then constituted. An important strategic goal of IGTN was to limit the WTO to trade issues and get it out of areas such as agriculture, government procurement and domestic regulation, where trade liberalization could have critically adverse effects on the livelihood of poor men and women. The question from the point of view of IGTN was: gender mainstreaming into what?

IGTN argued that there is a difference between seeking to engender trade analysis and policy, on the one hand, and gender mainstreaming the WTO, on the other hand. It saw value in engendering trade agreements, but mainstreaming gender into the governance and institutional structure of the WTO, at that particular historical juncture, was of questionable value. Engendering trade agreements was not simply about increasing the number of women in delegations or in the secretariat; even in the early years of the new century there was a significant number of women trade ministers, ambassadors and technical staff, though uneven across country delegations and within the WTO secretariat.[10] Rather, the problem was the nature of the ideology, technical analysis and political process that drove the organization and which underlie trade liberalization.[11]

Thus IGTN's imperative was to work to eliminate or mitigate the negative short- and long-term effects of WTO-led trade liberalization on women's lives and livelihood. The strategy was to develop and implement a critical political analysis approach to development and trade from a gender perspective. In doing so, IGTN sought to collaborate with global women's and other social movements on a transformative social and economic agenda. IGTN also focused on enhancing local and national capacity to undertake trade policy advocacy by training social and women's movements, parliamentarians and business and professional associations interested in promoting women's empowerment and gender equality.

A successful network?

Margaret Keck and Kathryn Sikkink identify five areas of influence of a transnational advocacy network, and probing these areas can help assess the achievements of IGTN. The five areas are: (1) issue creation and agenda setting; (2) influence on policy discourse of states, international organizations, NGOs and social movements; (3) influence on institutional procedures; (4) influence on policy change in target actors such as states, international organizations (international financial institutions, WTO, UN) or private actors (corporations); and (5) influence on state behavior (implementation of policy).[12]

Even with this helpful approach, given the vast mix of actors from academic researchers on gender and trade, to large development NGOs, to institutional players and committed government technical experts and decision-makers now involved in pushing a gender and trade approach, it is difficult to say what IGTN accomplished. What is clear is that from 2000 to 2007, IGTN was largely responsible for putting gender on the trade agenda of NGOs, popular movements, governments and multilateral institutions. It participated in the development of a gender and trade agenda and provided substantive analysis and responses to the actual trade negotiations. Unquestionably, IGTN became the "gender face" of the social movements working on trade and development and helped to shift the direction of the agenda from "trade is gender neutral to popular issue"[13]—so that with respect to Keck and Sikkink's first two areas of influence, IGTN can be said to be a successful network. In terms of phases 3, 4 and 5, influence on policy change and on the behavior of states, the evidence is less strong.

Since the early part of the new century, international institutions increasingly have turned their attention to gender and trade. UNIFEM was the first UN agency to focus on the issue with a publication on *Women and the New Trade Agenda*.[14] UNCTAD, which was one of the first agencies to host an expert group meeting on the subject of gender and trade, is, as noted above, now the focal point of the UN Inter Agency Task Force on Gender and Trade.[15]

The Commonwealth Secretariat has also been active in the field of gender and trade; it has organized three international seminars with policy makers, national women's machineries and trade negotiators on the subject. Its program on gender, trade and export promotion was a three-year program of capacity-building which brought together representatives from gender machineries, ministries of trade and export promotion departments along with national and regional institutions for a series of five-day training programs that addressed gender in trade and export management. The program, which built on the findings of IGTN researchers and involved an IGTN member global research coordinator as a key facilitator, was delivered in all the regions of the Commonwealth. As a result some official trade organs have integrated gender into their research and assessment frameworks. In addition, the Commonwealth Secretariat has produced a handbook on gender mainstreaming in the multilateral trading system, a training manual on gender, trade and export promotion and a volume of case studies on women, trade liberalization and export promotion.[16]

A landmark event was the 2003 WTO public symposium in which the director-general of the WTO was the keynote speaker at the plenary on "Women as Economic Players in Sustainable Development." For the first time a high-level official of the WTO acknowledged by his presence and in his remarks the importance of women's contribution to development and global trade. In addition, women ambassadors at the WTO also began to play a more active role to foster a breakthrough for gender equality concerns.

Finally, as noted by UNCTAD, "some countries are calling for specific gender assessments of trade agreements. For example, the African, Caribbean and Pacific Group of States (ACP)–EU Joint Parliamentary Assembly called upon the ACP States to undertake sustainability studies to inform the Economic Partnership Agreements (EPAs) negotiating process."[17] Some trade agreements have provisions that make reference to women and gender[18] while others—such as the North American Free Trade Agreement (NAFTA) and the Canada-Chile Free Trade Agreement—have side agreements that include gender.[19]

Despite this growing attention to issues of gender and trade in the work of international organizations, there are at least three weaknesses in the political process of incorporating a gender perspective into the multilateral trading system: First, the focus on gender and trade is still relatively small and often tends to be segregated within the larger programs of the institutions discussed above. For the most part, efforts to explore the linkage between gender and trade are maintained within the gender section of most institutions with not much leakage/linkage to the major economic analysis divisions or departments of these institutions. Second, trade negotiators tend to be resistant to the idea of gender seemingly on two grounds: Many do not understand gender and how it fits analytically into their negotiations mandate, and many see it as a Trojan horse for introducing policy conditionality into trade policy. But increasingly, this reticence is giving way to the reality that, in the words of one developing country trade official:

> the issue will not simply go away. It has validity. We have challenges with gender, so we have to be on the offensive and take it on before it is brought into the trade arena as conditionality that we have to fight against. So we are being proactive in examining the nature of the relationship and see how we can modify our trade policy and negotiations mandate to help mitigate the damages of trade liberalization on women workers etc.[20]

A third obstacle to the political acceptance of gender as a relevant dimension of trade is that gender machineries of various governments, with the exception of a few, are not well informed about the application of gender analysis to trade. Therefore they are not able to intervene in national trade policy decision-making even if there is political space to do so.

Lessons learnt from the IGTN experience

There are many lessons to be learnt from the IGTN experience. A few of the more strategic ones are highlighted below.

Lesson one: build on a firm base—political legitimacy

Trade negotiators and governmental decision-makers always want to know the source of legitimacy for them to incorporate what they perceive as "new" issues. IGTN, thus, at the beginning of any first intervention always sought to underscore that gender equality was not a "new" issue but rather a commitment emanating from agreements that governments have signed, including CEDAW, the Beijing Platform for Action and the output from the series of follow-up meetings. Further, most governments are also explicitly committed to social equity and social development objectives of the World Summit for Social Development and the evolving set of human rights instruments. More recently over 180 heads of state made a commitment to a set of Millennium Development Goals (MDGs) to be achieved by 2015. In all these agreements gender equality is recognized as a central concern. When IGTN worked with government officials and trade negotiators, CEDAW and the Beijing Platform for Action, key sources of reference with the provisions linking gender equality to the economy and macroeconomic policy were underlined in red.

Lesson two: holding responsible actors accountable

The first line for civil society participation in the governance of trade is to hold governments accountable at the local, national, regional, and global levels for their commitments to gender equality and gender mainstreaming. This means that gender analysis and perspective should be an integral part of the trade policy-making environment; there should be effective and meaningful coordination among sector ministries such as health, education, welfare, and agriculture; and there should be significant participation of national and regional gender machineries in the trade decision-making process and in developing gender policy and gender briefing documents aimed directly at influencing the direction of trade policies at the national, regional and multilateral levels.

However, the effective participation of gender machineries and women's groups in decision-making processes of the multilateral trading system requires clarity about the key issues being negotiated and how they are related to short-, medium- and long-term strategic gender interests. These groups therefore need to identify these issues and relationships in order to be able to hold governments accountable not only for domestic trade policy but also for the impacts of regional economic (and trade) arrangements to which they are parties.

Lesson three: build on a firm base—analytical legitimacy

It is pivotal to show the analytical grounds for integrating gender concerns into economic issue areas, especially from authoritative sources that governments and institutions recognize. IGTN legitimizes gender as an important variable in economics and economic policy analysis by referring to the work of the World Bank and feminist economists. The World Bank's publication on Engendering Development,[21] two volumes of the World Development Journal devoted to the subject matter,[22] and the gender chapter of the World Bank's Poverty Reduction Strategy Paper Sourcebook[23] have provided weight for leveraging the analysis on gender and trade. Many gender advocates are at odds with the bank on the wider issues of economic and social development and the overriding trade philosophy. But the bank certainly is a highly respected policy interlocutor in the economic and trade arena. So using its work is both necessary and sufficient for making the case with government officials that gender is an important variable in economic and trade analysis.

The IGTN model of a gender analytical approach to trade policy-making highlights three issues. First, it calls into question the notion of efficiency and gender-biased processes accompanying trade liberalization, such as commodification, privatization, commercialization and trade intensification, and assesses a country's trade strategy against economic and social development goals. Second, such an approach has the potential of generating broader and more effective policy targets and policy instruments, and of adding new content to the substantive agenda. This was the case, for example, in the agricultural negotiations on special products,[24] in the unearthing of deeply embedded equity issues in services liberalization, or in the Agreement on Trade Related Aspects of Intellectual Property Rights (TRIPS) discussion. Finally, a gender analytical approach adds to the voice and visibility of women and other underrepresented sectors of the economy. Hence, it widens the scope for participation, democracy and transparency.

Lesson four: engage gender perspectives and analysis in a critical and comprehensive way

The process of engendering and mainstreaming must always be seen as a political process that questions conventional wisdom and brings new insights to the challenges at hand. In the case of IGTN, the goal was to develop a critical perspective on trade policies by asking what the objectives of these policies were and whom they would serve. The second core imperative was to embed the understanding that gender analysis offers a more comprehensive, in-depth and deeply integrative approach to trade and development. As developed within the body of feminist political economy, gender analysis focuses on the interrelationship between the household sector, the productive sector and the informal sector. It seeks to pinpoint explicitly the transfer of costs and the burden of adjustments associated with trade liberalization on unpaid labor in the household economy, and the negative effects of this transfer on gender equality and women's and men's economic empowerment.

Ultimately, the goal is to get trade policy decision-makers to recognize and accept that good practices that will promote benefits for women from trade (liberalization) policy must start with a trade policy that is oriented towards poverty elimination and the promotion of gender equality as explicit objectives. It must be centered in a gender-aware model of the economy that takes account of the reproductive and informal sectors of the economy.

Lesson five: forward momentum—indicators and assessment tools

It is important to advocate strongly for more investment in the development of assessment methodologies for investigating the quantitative and qualitative economic, environmental and social impacts of trade agreements and policies. Examples of such assessment tools range from those that focus on the social and environmental effects of trade policy, generally referred to as Social Impact Assessments (SIAs, pioneered in the trade area by the European Commission, but not with an eye towards gender), to those that are focused on a paradigmatic approach to economic development such as the Human Development Impact Assessment (HDIA, developed by the UNDP), and Gender Trade Impact Assessment (GTIA, developed by Women's Edge).

GTIAs are distinct from HDIAs or SIAs in that they focus on the development of gender-aware indicators and modeling techniques for assessing the impacts of trade agreement provisions on gender equality and on men's and women's well-being. The framework for these assessments draws on concepts such as the sexual division of labor, labor market segmentation/discrimination, asymmetric rights and responsibilities, reproduction and motherhood, gendered social norms, time poverty, and political, economic, and social rights.

Conclusion

Gender and trade advocates are beginning to find their way—both analytically and empirically—in policy analysis and in the monitoring of the multilateral trading system. While there are many constraints and challenges, there are also numerous opportunities for deepening the work on gender and trade. Exploring these opportunities needs better coordination of information, consolidation of policy-oriented research and the proactive intervention of feminist scholars, women and gender machineries at the national, regional and global levels. UN Women, which includes UNIFEM, has continued the work on gender and trade, building on its comprehensive baseline, like multi-country research focused on tackling the opportunities, challenges and constraints facing women informal cross-border traders in the African regions.

The strategies utilized by IGTN have been important and necessary considering the context and conjuncture in which it operated. They have helped bring gender issues into the public policy arena, but have been less successful in mobilizing the full power of the women's movement. The political momentum for a critical analytical and policy approach is evaporating with the institutional takeover of the gender and trade agenda and its transformation into a much narrower and more instrumental agenda that legitimizes the trade system by "making trade work for women" rather than rethinking the trade policy environment itself. Therefore, more than ever, there is a need for continued IGTN global advocacy on gender and trade. As noted by Riley "[t]he political process of incorporating gender issues into trade policy is part of the transforming of current trade agenda beyond market access and efficiency toward trade that enhances human rights, social reproduction and social development."[25]

The conjecture of debts, deficits and austerity measures unleashed in rich and poor countries alike in the aftermath of the 2008 financial crisis presents multiple opportunities for grassroots women's organizations, policy analysts, academics, and—last but not least—UN Women to interrogate the new era of "prudential" macroeconomic policies against precautionary or pro-employment/anti-poverty approaches to macroeconomics and trade policies. There is also significant potential in the emerging stream of work linking macroeconomics and economic, social and cultural rights.[26] Since human rights have been a particularly effective grounding for women's organizations, it can be expected that many more women's rights organizations will be paying attention to macroeconomic, financial and trade policies.

Notes

1 The Informal Working Group on Gender and Trade includes Alternative Women in Development (Alt-WID, US), Association of African Women for Research and Development (AAWORD), Caribbean Association for Feminist Research and Action (CAFRA), Council for the Economic Empowerment of Women in Africa (CEEWA, Kenya), Catholic Institute for International Relations (CIIR), Educational International (EI), Gambia Women's Finance Association, International Coalition for Development Action (ICDA, Brussels), the International Confederation of Free Trade Unions (ICFTU), Oxfam-Ireland, Self Employed Women's Association (SEWA, Nepal), Network Women in Development Europe (WIDE), Women Working Worldwide (WWW) and the Women's Environment and Development Organization (WEDO).

2 This was the battle cry of the broad coalition of social movements and international NGOs organized against rampant trade liberalization. These groups drew attention to the negative impacts of trade agreements on livelihoods, employment and poverty. They also called for a halt in further trade liberalization until proper social and poverty impact assessments were undertaken of the outcomes of trade liberalization thus far. It was also a

call for correcting the unfairness, deficiencies and imbalances in trade rule-making that privileged the interests of developed countries and worked to the detriment of many developing countries. See: Bhagirath Las Das, *The WTO Agreements: Deficiencies, Imbalances and Required Changes* (London: Zed Books, 1998).

3 Food security exists when "all people, at all times, have physical and economic access to sufficient, safe and nutritious food that meets their dietary needs and food preferences for an active and healthy life" (World Food Summit, 1996). By this definition the critical factors are availability and access to food (e.g., through food imports). In contrast, food sovereignty is the right of peoples to healthy and culturally appropriate food produced through ecologically sound and sustainable agricultural practices.

4 Domestic regulation is part of the negotiations of the General Agreement on Trade in Services (GATS). It deals with topics such as public services, education and telecommunications.

5 For example, the Cotton-4 (Benin, Burkina Faso, Chad and Mali), the four cotton-producing countries in West Africa who work to get the United States to reduce cotton subsidies.

6 See for example: Maria Pia Hernandez, *Incorporating Gender Considerations for the Designation of Special Products in WTO Agriculture Negotiations* (Geneva: IGTN, 2005).

7 Susan P. Joekes and Ann Weston, *Women and the New Trade Agenda* (New York: UNIFEM, 1995).

8 UNCTAD, *Trade, Sustainable Development and Gender* (Geneva: UNCTAD, 1999); Anh-Nga Tran-Nguyen and Americo Beviglia Zampetti, ed., *Trade and Gender. Opportunities and Challenges for Developing Countries* (Geneva: UNCTAD, 2004).

9 See *Designing Gender Sensitive Foreign Trade Policy for India* sponsored by Ministry of Commerce on behalf of the Commonwealth Secretariat and *National Export Strategy: Gender Dimensions* (Government of Uganda).

10 Presently one of four deputy directors-general of the WTO is a woman, Valentine Sendanyoye Rugwabiza (since 2005). The WTO also presents gender disaggregated data on its staff complement on its website and argues that its total staff is equally balanced between men and women.

11 Gigi Francisco, *Why We Cannot Mainstream Gender into the WTO*, www.glow-boell.de/media/en/txt_rubrik.../Engender_WTO_Francisco.ppt.

12 Margaret Keck and Kathryn Sikkink, *Activists Beyond Borders: Advocacy Networks in International Politics* (Ithaca, NY: Cornell University Press, 1998).

13 Maria Riley, "Gender, Economics and Advocacy," presentation at the Symposium on "Gender and Trade Liberalization Global-local Processes and Gendered Impacts in Asia and Beyond," Southeast Asia Research Centre, City University of Hong Kong (22–23 August 2005).

14 Susan Joekes and Ann Weston, *Women and the New Trade Agenda* (New York: UNIFEM, 1995).

15 That includes: International Labour Organization (ILO), Food and Agriculture Organization (FAO), Economic Commission for Europe (ECE), Economic and Social Commission for Asia and the Pacific (ESCAP), Economic Commission for Latin America and the Caribbean (ECLAC), Economic and Social Commission for Western Asia (ECSWA) and European Economic Commission for Africa (ECA).

16 Mariama Williams, *Gender Mainstreaming in the Multilateral Trading System: A Handbook for Policy-makers and Other Stakeholders* (London: Commonwealth Secretariat, 2003); Catherine Atthill, Sarojini Ganju Thakur, Marilyn Carr and Mariama Williams, *Gender and Trade Action Guide* (London: Commonwealth Secretariat, 2007); Marilyn Carr and Mariama Williams, ed., *Trading Stories. Experiences with Gender and Trade* (London: Commonwealth Secretariat, 2010).

17 *Mainstreaming Gender in Trade Policy – Note by the UNCTAD Secretariat* (UNCTAD document TD/B/C.I/EM.2/2/Rev.1), 19 March 2009, 14.

18 See, for example, the Cotonou Agreement: preamble, article 1, article 9 (2) and article 20 (1) (2) state the parties' respect for international conventions regarding women's rights, their belief in men's and women's equality, and their commitment to include a gender perspective in "all areas of cooperation."

19 The side agreement of the American Free Trade Agreement on labor cooperation focuses on three issues: (a) the equality of women and men in the workplace; (b) the elimination of employment discrimination, including gender-based; and (c) equal pay for women and men.

20 Personal communications in debriefing at a country-level gender and trade capacity-building workshop.

21 World Bank, *Engendering Development Through Gender Equality in Rights, Resources, and Voice. World Bank Policy Research Report* (Oxford: Oxford University Press, 2001).

22 Special issues of *World Development* 23, no. 11 (November, 1995) and 28, no.7 (early summer 2000). The articles in this journal were primarily written by members of the International Working Group on Gender, Macroeconomics and International Economics (GEM-IWG).

23 Jeni Klugman, ed., *A Sourcebook for Poverty Reduction Strategies* (Washington, DC: World Bank, 2002).

24 These are products exempted from tariff reductions. At one point in the WTO discussion around these some country delegations investigated whether gender could be condition for classifying such products as products critical to women's livelihood in a particular country.

25 Riley, "Gender, Economics and Advocacy."

26 Rhadika Balakrishnan and Diane Elson, *Rethinking Macroeconomic Strategies from a Human Rights Perspective* (New York: Marymount Manhattan College, 2009).

Part II

Human rights of women

6 Translating international women's rights norms

CEDAW in context

Susanne Zwingel

- **Theorizing norm translation: a transnational view on international women's rights**
- **CEDAW in global discourse translation**
- **Conclusion: lessons from feminist strategies in global discourse translation**

Over the last decades, gender equality norms have become an integral part of global governance institutions. Departing from this international success story—most recently manifested in the creation of the United Nations Entity for Gender Equality and the Empowerment of Women (UN Women)—activists and scholars alike have been trying to figure out if the engendering of international institutions has translated into concrete legal, political and socio-economic change to the benefit of the world's women. In a nutshell, the question is if international gender equality norms radiate into the world and how and under which conditions this is most likely to happen. This contribution focuses on one piece of international governance to investigate this question, namely the Convention on the Elimination of All Forms of Discrimination against Women (CEDAW or Women's Rights Convention), because it has been the cornerstone of the international women's rights discourse for over 30 years and because the concreteness of its procedure offers itself for an actor- and context-specific analysis.

The chapter consists of three parts: It first provides a theoretical framework that helps to understand processes of norm translation. Contrary to the mainstream of norm diffusion literature that often takes a one-directional look at norms as globally produced and domestically implemented, norm translation is conceptualized as an open process of negotiation in which various actors in international, transnational, national and local contexts are involved. Accordingly, the framework developed here identifies three relevant dynamics of "traveling norms" which I call global discourse translation, impact translation, and distorted translation. The following empirical section focuses on the first dynamic—global discourse translation—in respect to the CEDAW Convention: It gives an overview of CEDAW's creation, the functioning of its state-centered monitoring body, and the contribution of non-state actors to its monitoring procedure. I concentrate on the dynamic of global discourse translation here because on the one hand, it shows the "incompleteness" or ongoing contention regarding the content and legitimacy of gender equality norms on the international level. On the other hand, it is a prime case study of the feminist strategy of persuasion and connecting contexts. The concluding section discusses implications of my findings for strategies of norm translation reaching beyond the international level.[1]

Theorizing norm translation: a transnational view on international women's rights

Most International Relations literature on human rights norms focuses on the ways in which global norms are created and implemented, and on the factors that inhibit or enhance implementation. Typically, a logic of cause and effect is implied, where international norms are imagined to cause domestic effects. Regime analysts usually assume a linear if complex process of regime formation, usually finalized with treaty adoption, and have diverging expectations regarding regime implementation: Constructivists underline the opportunity for states to create a non-coercive platform of cooperation and joint problem solving;[2] Neoliberals do not expect states to implement internationally agreed upon norms if they do not have incentives to do so.[3] Literature on global norm diffusion goes beyond the analysis of treaties and explores more broadly how norms emerge on the international level and then become meaningful in domestic contexts. For example, Finnemore and Sikkink developed a "norm life-cycle" model starting with global norm emergence leading to cascading and ending in domestic norm internalization, and Keck and Sikkink place emphasis on transnational advocacy networks as powerful "norm entrepreneurs" vis-à-vis governments that block the implementation of international norms.[4] The power of such rather weak actors comes from their ability to create and disseminate credible information that puts pressure on governments to act.

Despite important insights, the approaches presented so far rest on a number of problematic assumptions: First, human rights and international norms in general are considered unequivocally defined. Second, it is assumed that the world can be divided into a norm-abiding international community of (liberal) states and a number of deviant (authoritarian) states and that the second group needs to be socialized into desirable behavior. Finally, global norm diffusion literature constructs nongovernmental actors as altruistic norm advocates vis-à-vis self-interested norm-violating governments. Instead, I propose to conceptualize international norms as open-ended, evolving discourses, not all dimensions of which are equally accepted. I suggest that states as those responsible for implementing global norms represent a mix of divergent value systems that all have the potential to overlap with and differ from international norms—in the field of women's rights, this would mean to replace a dichotomous categorization of "progressive" vs. "backward" states with the understanding that women's rights are subject to a wide range of interpretations and that all states deviate from the ideal of gender equality, if to varying degrees. Finally, I argue that transnational networks have to be understood as groups of actors that have their own political agenda and are not automatically knowledgeable and supportive of international norms.

In order to come to a less global-centrist view that I consider necessary for a sound understanding of norm translation, it is useful to draw on the interdisciplinary field of transnationalism studies that focuses on "different forms of rearrangements of geographic-social spaces beyond, alongside and above the formerly dominant national society paradigm."[5] Three elements from this literature seem crucial for theorizing norm translation: First, transnationalization literature moves away from the discourses of separation—the global as linked to abstraction and power, the local as linked to the concrete and limited—and describes a new understanding of the socio-geographic interrelatedness of global, national, and local spheres.[6] Applied to CEDAW, such mutually constitutive situatedness suggests a transnational assemblage between the international set of norms that CEDAW represents, the national apparatus that has made a commitment to implement it, and the everyday life situations in which women's rights are supposed to work. Second, the homogenizing influence of globalization—or global norms—is questioned. Hannerz argues that the power of globalization is always confronted with strong social sites of everyday life that have the potential of contestation.[7] In Massey's view, any space is under constant negotiation by those who inhabit and shape a locality. For a de-centered analysis of international norms, this idea of "the local" suggests a variety of context-dependent norm implementation processes. It also implies the notion of self-determination: no idea that has been agreed upon "somewhere" should be closed to further negotiation in the context where it is supposed to be applied; rather, it should serve to open—or add to—normative debates.[8] Third, transnationalization literature reminds us that the idea of the local as culturally bound and unchanging is misleading. To the contrary, transnational influences can be identified as always existing forces in the creation of culture.[9] This understanding is helpful to overcome a dividing notion of culture as it is often constructed in the field of women's rights, e.g. when culture is identified as static, backward, and the main source of women's rights violations.[10] Instead, it is more appropriate to understand the Convention itself, the international monitoring procedure, and all sorts of situated gendered practices as culturally produced positions.

Feminist literature on transnationalism shares the presented assumptions yet has a focus on the manifold forms of border-crossing women's activism and contextualized gender orders.[11] Within transnational cultural flows, gender relations are shaped by "scattered hegemonies" simultaneously consisting of "global economic structures, patriarchal nationalisms, 'authentic' forms of tradition, local structures of domination, and legal-juridical oppression on multiple levels."[12] To use this feminist transnational lens for an analysis of CEDAW repercussions means to de-homogenize, de-essentialize and, basically, complicate agency positions, in particular those of international institutions, states, and nongovernmental organizations (NGOs). *International institutions* do partly represent a consistent set of values, but often also send ambivalent normative signals to states, for example if structural adjustment programs and socio-economic human rights are supposed to be implemented simultaneously. However, CEDAW as a state-centered women's rights mechanism is only equipped to address gender hierarchies that are domestically produced, not the ones that result from international free trade regimes[13]—it is thus part of an inconsistent international normative framework. Likewise, *states* have to be understood as historically contextualized entities embodying differing levels of factual sovereignty and a broad variety of attitudes regarding international gender norms. States may both perpetuate hierarchical gender orders and/or figure as important promoters of gender equality, in particular because of the key role states play in regulating family relations and reproductive rights.[14] Finally, *NGOs* are often altruistic norm advocates that put moral pressure on powerful state actors, but that is not all they can be: in particular if NGOs are internationally funded, they may be (perceived to be) part of a hegemonic international coalition that imposes certain values on those they are supposed to help;[15] they may contribute to the fragmentation and disempowerment of domestic movements;[16] and they may form precisely to resist international norms and foster a culture of domestic or transnational counter-hegemony.[17]

The re-construction of international norms as components of a semi-coherent "web of meaning," states as entities with varying forms of sovereignty and gender regimes, and NGOs as norm entrepreneurs driven by a variety of motives sets the stage for analyzing the dynamics of "traveling[18] gender norms." I propose three processes as particularly relevant: Global discourse translation, impact translation and distorted translation. I use the term "translation" because it allows different avenues of cross-cultural encounters and transmissions of meaning, global to national, local to global, etc., yet the term also includes unevenness—mutual enrichment is possible as well as manipulation.[19] The first process—*global discourse translation*—describes activism out of different contexts to influence international institutions, or activists' use of such institutions to create transnational links and strategies of action.[20] It implies that a norm or discourse is not just "placed" on an international agenda, but that it remains subject to interpretation, adaptation, restriction or expansion—in other words, global discourse translation is unfinished and ongoing. The second dynamic—*impact translation*—means agency that uses international gender norms to influence domestic gender regimes and is situated in and complicated by multiple other normative and material settings. Studies that have analyzed this formation place emphasis on thoroughly characterizing both the actor constellation and the context relevant for the action of translation, as this context often determines strategies and outcomes.[21] Levitt and Merry have coined the term "vernacularization" to describe various (more or less transformative) ways of making international gender norms culturally understandable and acceptable.[22] The third constellation—*distorted translation*—refers either to transnational activism in which international norms play only an indirect role or to a more obvious disconnect between international and domestic norms. For example, international gender norms have been found to be much less significant for public gender policy formation in many post-industrialized democracies than domestic actor constellations,[23] and in a number of states, it adds legitimacy to women's activists' claims if they do not refer to international gender norms because of strongly diverging hegemonic gender norms.[24] All three dynamics are not clearly separable from each other—impact translation and distorted translation differ only gradually, and global discourse translation may be intertwined with both.

CEDAW in global discourse translation

CEDAW was adopted by the United Nations (UN) General Assembly in 1979 and entered into force in 1981, in the middle of the UN-proclaimed

Women's Decade (1976–85). Due to its comprehensiveness and legally binding character, the convention has often been called a milestone for the world's women.[25] At its 30th birthday in December 2009, UN Secretary-General Ban Ki-Moon called CEDAW "one of the most successful human rights treaties ever."[26] This section presents the dynamics of global discourse translation around CEDAW. The creation of the convention and the work and development of the convention's monitoring mechanism, including the contribution of non-state actors, show that CEDAW, once adopted, was not just an international set of norms to be domestically implemented. Instead, it has been further developed and interpreted within the international context itself. Following the description, I discuss three noteworthy dynamics in this process, namely the gap between rhetorical and substantial support of women's rights norms, the identity construction of the CEDAW Committee, and the continuing wresting over the most adequate interpretation of what is "right" for women.

The global context in which women's rights norms have emerged and developed can roughly be divided into three broad phases. In the first phase of *international knowledge formation* from 1945 to 1974, an understanding of the far-reaching repercussions of gender hierarchies was created, yet these insights remained largely rhetorical and did not trigger significant change. In a second phase of *global transformation* from 1975 to 1995, gender equality issues were catapulted into the limelight of international awareness, both in terms of the publicity that gender equality received, the far-reaching institutional changes within the UN, and the network formations of what came to be called a global women's movement.[27] The last phase from 1996 to the present, the phase of *consolidation and shift of allegiances*, is shaped by disenchantment regarding the stagnating or even worsening socio-economic conditions many women live in, despite global and governmental rhetoric, but also by continued institutional strengthening of gender awareness.[28]

The drafting of CEDAW falls into the second phase of this scheme. It is telling that increased international sensitivity towards gender equality was needed to create CEDAW, despite the fact that the idea of non-discrimination on grounds of sex had technically been an integral part of the human rights framework from its inception in 1948. Feminist lawyers have provided in-depth critiques of this gender blindness in human rights law that is, among other things, rooted in the conceptual exclusion of the private sphere.[29] As the first comprehensive manifestation of gender-sensitivity in international human rights law, CEDAW is shaped by an understanding of gender equality that aims at eliminating all forms of discrimination against women in order to achieve a situation of equality between men and women. While this notion was criticized as being unable to transform the existing system according to women's needs,[30] CEDAW was seen by many as "the definitive international legal instrument requiring respect for, and observance of, the human rights of women."[31] Despite controversies in the drafting process regarding the understanding of women as either individuals or role-players in family and society, the final document envisions women as autonomous rights holders.[32] The convention's monitoring mechanism is designed cautiously and is exclusively state-oriented. What is striking about CEDAW is the stark contrast between strong ideals on the one hand—among others, the comprehensive definition of discrimination; the attempt to overcome gender stereotypes and the resulting gendered division of labor, both based on female inferiority; the focus on equality of outcome—and halfhearted enforcement mechanisms on the other, for example, vague formulations in respect to state obligations and a uniquely short annual meeting time for the CEDAW committee.[33] This contrast seems to represent two types of drafters: women's rights advocates and representatives of states' interests.

The convention entered into force on 3 September 1981, after the ratification of the 20th state. In the following year, the CEDAW monitoring committee started to convene to oversee the implementation of the convention. The committee started as a weak, almost unknown human rights instrument that developed into a visible, well-known and authoritative treaty body. Four dimensions are important to understand this metamorphosis: First, the CEDAW committee received inadequate administrative support in its early years. It was served by a part of the UN Secretariat that dealt with all women's concerns within the UN, was understaffed and not sufficiently knowledgeable in human rights matters. The committee was also restricted in terms of its meeting time—to the extent that it could not carry out even its basic functions.[34] Since the mid-1990s, the committee's working conditions have improved on all levels: it has developed administrative service that is highly competent in the field of women's rights, initially from the Division for the Advancement of Women and, since 2008, from the Office of the High Commissioner for Human Rights, the secretariat that serves all human rights treaty bodies.[35] Further, its meeting time was gradually expanded from two to nine weeks annually.

Second, the committee has consistently developed its style and working methods. Its position in the constructive dialogue is to support states parties[36] to the convention in the fulfillment of their treaty obligations. It voices criticism, but usually strives for a non-adversarial atmosphere in order to encourage the willingness of states parties to accept scrutiny. It tries to be aware of the different starting points and cultural contexts of the states parties, but aims in each dialogue to enhance the respective state's compliance with the convention. There is agreement among committee members that the obligations laid out in the convention are the same for all, yet meaningful context-specific measures to fulfill the obligations may differ. Beyond this general position, the committee has worked to increase its efficiency, for example in thoroughly preparing and structuring the dialogue, in systematically considering information from independent sources, and in issuing concluding observations that contain clear proposals as to how a specific state should fulfill its treaty obligations. Further, the committee has, both in its statements in regard to potentially inadmissible reservations[37] and in its comprehensive general recommendations, demonstrated a self-assured position as the authoritative body interpreting the convention, for example by including concerns that are not explicitly mentioned in the original treaty such as violence against women and intersectional discrimination.

Third, the convention's and committee's international standing went from a rather unknown mechanism to one with significant intra-institutional relationships. Partly due to the commitment of committee experts and UN staff members, partly due to the increasing recognition of women's rights generally speaking, other UN agencies started to take notice of the convention and to interact with the committee. Of particular relevance both for CEDAW's recognition and its application is the interaction with the other human rights treaty bodies and the United Nations Development Fund for Women (UNIFEM, now part of UN Women). The fourth and final dimension of CEDAW's improved international standing is the creation of its Optional Protocol (OP) in the year 2000 that has enabled the committee to take on a new dimension of women's rights monitoring, namely dealing with individual complaints and inquiry procedures. In the decade since the OP's entry into force, the committee has dealt with fewer than 20 cases. In general, the committee's practice under the OP has shown that it aims at straddling a tension between legal accuracy—the "currency" of its legitimacy among states—and a decidedly victim-oriented reading of the convention.[38]

A final dimension of CEDAW in global discourse translation lies in the contribution of NGOs. While NGOs are not built into the state-centered construction of the CEDAW monitoring process, they have turned out to be pivotal for the translation of the norms enshrined in CEDAW. The two relevant NGO interventions regarding global discourse translation are, one the one hand, NGOs that support CEDAW and work directly with the committee. One the other hand, there are women's organizations that aim at delegitimizing the women's rights approach on the international level; they hardly ever directly influence the CEDAW monitoring procedure and they are less organized than the supportive NGOs, but they have been gaining strength.

NGOs supportive of the CEDAW process have prepared independent information for the monitoring procedure against which the governmental reports can be measured—this has made the dialogue more comprehensive and demanding for governments; they have made the convention known among women's NGOs in various countries and thus multiplied NGO input to the committee's sessions; and they have helped to shape the global discourse on women's rights in various forums. The NGO *International Women's Rights Action Watch Asia Pacific* (IWRAW AP) has been a motor for all these functions since its foundation in the early 1990s. It has brought women's NGOs from more than 120 countries to committee meetings when their government was reporting, conducted countless workshops on CEDAW and has been an outspoken voice in furthering the understanding of core concepts of women's rights. For example, IWRAW AP has contributed to the committee's integration of the concept of intersectional discrimination into some of its general recommendations and to its official acknowledgement[39] that discrimination based on sexual orientation is part of the mandate of the convention.[40]

Increasingly, organizations that style themselves as "pro-women" and "anti-feminist" have articulated positions at UN meetings that question the women's rights framework as represented by CEDAW. Generally speaking, these positions contradict the idea that women should be endowed with enforceable rights as individuals, in all fields, and on the same level as men—this is considered a radical feminist ideal. Instead, they emphasize complementary gender difference and the important role women play in the family and, thus, in society as a whole. Most fundamentally, pro-women/anti-feminist voices reject the notion of reproductive and sexuality rights because they consider them antithetical to the dignity of (heterosexual) motherhood and to the institution of the "natural family."[41] If pro-women/anti-feminist voices make reference to CEDAW directly, they mostly criticize the committee for its alleged defamation of motherhood, its pro-abortion stance and its support for the decriminalization of women's sexual orientation;[42] they do not reveal that parts of their agenda also overlap with the women's rights framework, expressed, for example, in a commitment to broadening women's career options.[43] Considering that the CEDAW committee had to wrestle for intergovernmental recognition, it is not surprising that committee experts do not want to be labeled "radical feminists," clearly a derogatory term in the UN context. Thus, pro-women/anti-feminist voices indirectly contribute to a more moderate tone of the committee, in particular in contested areas, even if individual experts may deviate from that trend.

Overall, this narrative of the convention's creation, the functioning of its international monitoring body, and the discussions it has triggered among supporters and opponents of the women's rights framework describes a process of growing rhetorical and material recognition, including contestation, on the international level. Three aspects of this case seem particularly relevant to understanding global discourse translation, namely the relationship between rhetorical and material intergovernmental support of women's rights norms, the identity construction of the monitoring body as authoritative in interpreting these norms, and the ongoing wresting over the most adequate interpretation of what is "right" for women.

As for the first aspect, the gap between the adoption of CEDAW and the long time it took before it was taken seriously as a human rights instrument both by states and within the UN seems, to say the least, puzzling. CEDAW was a crucial achievement of the global transformation phase from 1975 to 1995. However, during this phase of high awareness regarding gender equality, the CEDAW committee was largely invisible and had to confront all kinds of limitations. Only much later, from the mid-1990s on, did it unfold its full potential as a women's rights instrument. This time lag indicates that a lot of effort is necessary to turn a rhetorically accepted idea into practice, even on the international level only. Indeed, committee members and other CEDAW supporters have worked hard to make the convention relevant among other international players and states. Such constant international "norm work" is indispensable—if it is not undertaken, a norm that has become globally pervasive on a rhetorical level may never move into a stage where it receives substantial support. This finding underlines the notion of international norm creation as an open-ended experiment.

Tied to the first aspect is a second one, namely the way the committee has been constructing its collective persona to be considered an authoritative and legitimate interpreter of women's rights. One source of legitimacy is, according to convention article 17.1, committee members' competence in the field of women's rights. This expertise is heavily based on academic knowledge.[44] The experts can be described as a privileged, often times government-employed group consisting mostly of women. Also following article 17, the experts represent all world regions and different forms of civilization and legal systems. Notwithstanding the fact that this mixture excludes certain forms of competence—non-academic knowledge, among others—the committee understands itself as culturally inclusive. It considers this identity as suitable to work for the realization of the convention in a comprehensive manner. However, such diversity may lead to assessments of state practices that are more shaped by the priorities of the sum of individual experts than by one coherent standard.[45] As one expert puts it, "we speak with 23 voices. … What unites us is the text of the treaty and our commitment to human rights. But from different cultural, from different legal perspectives. … So when we address delegations, we … recognize that each one of us will put different accents, but always from within the parameters of the convention."[46] While cultural inclusiveness is one way to create legitimacy, committee members have also been juggling with the tension between resolute women's rights advocacy and the respect for state sovereignty. Sometimes, the concern for the latter has made the committee's position more cautious than necessary. In a paradoxical way, the committee's increased effort to enhance implementation of the convention is part of this dilemma. The increased concern of the committee to get more proactive in regard to implementation could lead to a less holistic vision of women's rights. In the words of an NGO representative, "they [the committee members] say … let's talk about specifics and implementation. Those two things cannot be the things that drive the committee. … The committee has to constantly raise the bar on women's rights. That is their first mandate. They don't implement, we implement."[47] Thus, the collective identity of the committee is one of culturally diverse women's rights experts, even if this identity may be less inclusive than the experts think. The committee has to consistently ponder how much women's rights advocacy is considered legitimate within an intergovernmental setting. This dilemma bears the risk to abandon its lighthouse function regarding progressive women's rights interpretation; it also shows, again, the unfinished journey of global discourse translation.

Finally, the consideration of nongovernmental contributions to CEDAW's global creation has opened another layer of contested legitimacy: NGO voices in support of a fundamental and undivided commitment to the women's rights framework have added weight to the committee's role as a women's rights advocate. However, the more articulate this position of women's rights advocacy has become, the more women's organizations with "anti-feminist" viewpoints have appeared on the international stage in order to strengthen a set of values that partly clashes and partly overlaps with the women's rights framework. This has, on the one hand, had a moderating effect on the committee's dealing with explosive issues; but we see on the other hand that the critics do not deny the legitimacy of a large part of the women's rights framework. These processes resemble earlier disputes, for example the fighting over (and eventually, agreement on) adequate terms during the drafting phase of the convention. This oscillation of positions indicates, once more, that the global debate on the meaning of women's rights cannot be settled with the adoption of a legally binding instrument such as CEDAW. Indeed, this is and will be an ongoing debate.

Conclusion: lessons from feminist strategies in global discourse translation

As we have seen, CEDAW represents a case of overall successful global discourse translation, yet this process has been complex, sometimes cumbersome, and is likely to remain open-ended. Two strategies employed by feminist actors within and outside the committee have been key in making CEDAW authoritative on the international level: One is the continuous effort to connect disconnected contexts, for example the rapprochement of discourses on gender equality and human rights within the UN or the integration of nongovernmental input into the purely state-centered monitoring procedure provided for by the convention. The other strategy is the struggle to find a balance between competing demands, visible in the committee's self-construction as a legitimate body in the eyes of states parties, expressed in a specific concept of expertise, cultural inclusiveness and a not-too-radical feminist agenda, that at the same time "raises the bar on women's rights" and is the centerpiece of global women's rights advocacy. In particular the first effort is likely to be enhanced by UN Women, as it is part of the agency's mandate to create intra-organizational coherence in regard to women's rights and gender equality. Likewise, UN Women will be benefitting from applying the women's rights framework that CEDAW represents to its manifold tasks including, following the practice of UNIFEM, to its operational work in domestic contexts.

While this contribution has focused on the complexities of global norm translation, the findings presented make at least three suggestions for norm translation processes beyond the international. First, it is reasonable to expect more diverse positions in regard to women's rights in domestic contexts than on the intergovernmental level—for example, positions openly rejecting women's rights or positions that are hard to link to the idea of individual dignity inherent in the human rights framework. Second, it is likely that the legitimacy of a position is not so much produced by expert knowledge and cultural diversity, but by political, cultural or religious idiosyncrasies and grown institutions, many of which may not be susceptible to women's rights. This is not to say that such forces do not play a role on the intergovernmental level, but rather that they play a greater role whenever a norm requires material and not only rhetorical support. As we have seen before it was easier to convince governments to ratify the convention than to have them set aside sufficient resources to make its monitoring mechanism effective. Finally, the way international women's rights norms connect with otherwise contextualized norms will fundamentally depend on transnational norm translators and their strategies to make international norms pervasive and accessible. I would argue that while it can never be guaranteed that international gender equality norms are embraced, it is nevertheless preferable to negotiate the meaning and usefulness of these norms in context rather than consider the international framework as a priori superior. Ideally, UN Women's operational mandate could enhance this translation process.

Notes

1 For a comprehensive analysis of processes of impact translation and distorted translation of CEDAW, see Susanne Zwingel, *Translating International Women's Rights: The CEDAW Convention in Context* (Basingstoke: Palgrave Macmillan, forthcoming).
2 Abraham Chayes and Antonia Handler Chayes, *The New Sovereignty. Compliance within International Regulatory Agreements* (Cambridge, MA and London: Harvard University Press, 2005).
3 Emilie H. Hafner-Burton, Kiyoteru Tsutsui and John W. Meyer, "International Human Rights Law and the Politics of Legitimation. Repressive States and Human Rights Treaties," *International Sociology* 23, no. 1 (2008): 115–41.
4 Martha Finnemore and Kathryn Sikkink, "International Norm Dynamics and Political Change," *International Organization* 52, no. 4 (1998): 887–917; Margaret E. Keck and Kathryn Sikkink, *Activists beyond Borders. Advocacy Networks in International Politics* (Ithaca, NY: Cornell University Press, 1998).
5 Ludger Pries, "Transnationalism and Migration: New Challenges for the Social Sciences and Education," in *Migration, Education and Change*, ed. Sigrid Luchtenberg (New York: Routledge, 2004), 45–69.
6 Doreen Massey, *For Space* (London: Sage, 2005).
7 Ulf Hannerz, *Transnational Connections. Culture, Peoples, Places* (New York: Routledge, 1996).
8 Massey, *For Space*.
9 Amartya Sen, *Development as Freedom* (New York: Oxford University Press, 1999), 242–44.
10 Sally Engle Merry, "Constructing a Global Law—Violence against Women and the Human Rights System," *Law and Social Inquiry* 28 (2003): 941–77.
11 Nancy A. Naples and Manisha Desai, ed., *Women's Activism and Globalization. Linking Local Struggles and Transnational Politics* (New York: Routledge, 2002).
12 Inderpal Grewal and Caren Kaplan, "Introduction: Transnational Feminist Practices and Questions of Postmodernity," in *Scattered Hegemonies.*

Postmodernity and Transnational Feminist Practices, ed. Inderpal Grewal and Caren Kaplan (Minneapolis, MN: University of Minnesota Press, 1994), 1–33.

13 Uché U. Ewelukwa, "Centuries of Globalization; Centuries of Exclusion. African Women, Human Rights, and the 'New' International Trade Regime," *Berkeley Journal of Gender, Law and Justice* 20 (2005): 75–149.

14 Lynn Savery, *Engendering the State. The International Diffusion of Women's Human Rights* (New York: Routledge, 2007); Valentine M. Moghadam, *Globalizing Women. Transnational Feminist Networks* (Baltimore, MD: Johns Hopkins University Press, 2005).

15 Kim D. Reiman, "A View from the Top: International Politics, Norms, and the Worldwide Growth of NGOs," *International Studies Quarterly* 50, no. 1 (2006): 45–67.

16 Sonia E. Alvarez, "Translating the Global. Effects of Transnational Organizing on Local Feminist Discourses and Practices in Latin America," *Meridians: Feminism, Race, Transnationalism* 1, no. 1 (2000): 29–67.

17 Doris E. Buss, "The Christian Right, Globalization, and the 'Natural Family,'" in *Gods, Guns, and Globalization. Religious Radicalism and the International Political Economy*, ed. Marry Ann Tétreault and Robert A. Denemark (Boulder, CO: Lynne Rienner, 2004), 57–77; Niamh Reilly, *Women's Human Rights: Seeking Gender Justice in a Globalising Age* (Cambridge: Polity Press, 2009), 140 ff.

18 The term "travel" is used to underline the notion of norms as non-static, not completed or arrived, but continuously exposed to new influences. I borrow this term from James Clifford, *Routes: Travel and Translation in the Late 20th Century* (Cambridge, MA: Harvard University Press, 1997), 2.

19 Paula G. Rubel and Abraham Rossman, ed., *Translating Cultures. Perspectives on Translation and Anthropology* (New York: Berg Publishers, 2003).

20 Peggy Antrobus, *The Global Women's Movement. Origins, Issues and Strategies* (London: Zed Books, 2004); Devaki Jain, *Women, Development, and the UN. A Sixty-Year Quest for Equality and Justice* (Indianapolis, IN: Indiana University Press, 2005); Reilly, *Women's Human Rights.*

21 Millie Thayer, *Making Transnational Feminism. Rural Women, NGO Activists, and Northern Donors in Brazil* (New York: Routledge, 2010); Anita Weiss, "Interpreting Islam and Women's Rights: Implementing CEDAW in Pakistan," *International Sociology* 18, no. 3 (2003): 581–601.

22 Peggy Levitt and Sally Merry, "Vernacularization on the Ground: Local Uses of Global Women's Rights in Peru, China, India and the United States," *Global Networks* 9, no. 4 (2009): 441–61.

23 Dorothy E. McBride and Amy G. Mazur, "Measuring Feminist Mobilization: Cross-National Convergences and Transnational Networks in Western Europe," in *Global Feminism. Transnational Women's Activism, Organizing, and Human Rights*, ed. Myra Marx Ferree and Aili Mari Tripp (New York: New York University Press, 2006), 219–46.

24 Rebecca Foley, "Muslim Women's Challenge to Islamic Law. The Case of Malaysia," *International Feminist Journal of Politics* 6, no. 1 (2004): 53–84; Benjamin Stachursky, "The Promise and Perils of Transnationalization. A Critical Assessment of the Role of NGO Activism in the Socialisation of Women's Human Rights in Egypt and Iran." PhD diss., Universität Potsdam 2010.

25 Charlotte Bunch, "Women's Rights as Human Rights: Toward a Re-Vision of Human Rights," *Human Rights Quarterly* 12, no. 4 (1990): 486–98; Rebecca J. Cook, "Women," in *United Nations Legal Order, Vol. 1*, ed. Oscar Schachter and Christopher C. Joyner (Cambridge: Cambridge University Press, 1995), 433–71.

26 Inter Press Service 2009, *Women's Treaty a Powerful Force for Equality*, http://www.ipsnews.net/news.asp?idnews=49545.

27 Elisabeth J. Friedman, "Women's Human Rights: The Emergence of a Movement," in *Women's Rights, Human Rights: International Feminist Perspectives*, ed. Julie Peters and Andrea Wolper (New York: Routledge, 1995), 18–35; Moghadam, *Globalizing Women.*

28 For more detailed accounts, see Antrobus, *The Global Women's Movement*; Jain, *Women, Development, and the UN*; Moghadam, *Globalizing Women*; and the introduction of this volume.

29 For example, Bunch, "Women's Rights as Human Rights," 486–98; Christine M. Chinkin, "Women's Rights as Human Rights under International Law," in *Understanding Human Rights*, ed. Conor Gearty and Adam Tomkins (London: Mansell Publishing, 1996), 553–72; Cook, "Women," 433–71.

30 Hillary Charlesworth, "What are 'Women's International Human Rights'?," in *Human Rights of Women. National and International Perspectives*, ed. Rebecca J. Cook (Philadelphia, PA: University of Pennsylvania Press, 1994), 58–84.

31 Cook, "Women," 441.

32 Lars Adam Rehof, *Guide to the Travaux Préparatoires of the United Nations Convention on the Elimination of All Forms of Discrimination against Women* (Dordrecht, The Netherlands: Martinus Nijhoff Publishers, 1993).

33 The complete text of the convention is available at http://www2.ohchr.org/english/bodies/cedaw/index.htm.

34 For a harsh critique of the committee's working conditions, see Andrew C. Byrnes, "The 'Other' Human Rights Treaty Body. The Work of the Committee on the Elimination of Discrimination Against Women," *Yale Journal of International Law* 14 (1989): 1–67.

35 A list of all human rights treaty bodies can be found at the website of the UN High Commissioner for Human Rights at http://www.ohchr.org.

36 'States party' is the correct legal term referring to any state that has ratified the convention and, thus, has become a party to it.

37 CEDAW article 28.2 allows ratification with reservations, but only if such reservations are not "incompatible with the object and purpose of the present Convention." However, the text does not specify which body is authorized to determine inadmissibility. See in more detail Rebecca J. Cook, "Reservations to the Convention on the Elimination of All Forms of Discrimination Against Women," *Virginia Journal of International Law* 30 (1990): 643–716.

38 Zwingel, *Translating International Women's Rights.*

39 In its General Recommendation no. 28 on article 2 of the convention, the committee spells out the obligations states parties have under the convention to eliminate *all forms* of discrimination against women. It states: "The discrimination of women based on sex and gender is inextricably linked with other factors that affect women, such as race, ethnicity, religion or

belief, health, status, age, class, caste and sexual orientation and gender identity." Committee on the Elimination of Discrimination Against Women, *General Recommendation No. 28 on the Core Obligations of States Parties under Article 2 of the Convention on the Elimination of All Forms of Discrimination Against Women* (CEDAW document CEDAW/C/GC/28), 16 December 2010, para. 18.

40 IWRAW Asia Pacific, *Possible Elements for Inclusion in a General Recommendation on Article 2 of CEDAW, Outcome Document of the Expert Group Meeting on CEDAW Article 2: National and International Dimensions of State Obligation* (Kuala Lumpur, Malaysia, 2007), http://www.iwraw-ap.org/aboutus/pdf/Elements_paper_final_version_Jan9.pdf.

41 Buss, "The Christian Right, Globalization, and the 'Natural Family,'" 57–77.

42 Wendy Wright, Concerned Women for America, *CEDAW Committee Rulings*, http://www.cwfa.org/articledisplay.asp?id=1870.

43 Real Women of Canada, *Our View, Canada's New Women's Movement*, http://www.realwomenca.com/page/ourviews.html.

44 Most committee members have a university degree; in distinction from other treaty bodies where the experts are usually lawyers, CEDAW experts are trained in diverse fields including literature, law, medicine, and social sciences.

45 Occasionally it becomes apparent that committee experts do not always agree on how to interpret the provisions of the convention. For example, Morvai makes it very clear that her anti-prostitution stance was not shared by other committee experts, yet she nevertheless stressed it consistently in the constructive dialogue. See Krisztina Morvai, "Personal Reflection: Rethinking Prostitution and Trafficking," in *The Circle of Empowerment. Twenty-five Years of the UN Committee on the Elimination of Discrimination Against Women,* ed. Hanna Beate Schoepp-Schilling and Cees Flinterman (New York: The Feminist Press, 2007), 141–45.

46 Interview with committee expert, January 2010.

47 Interview with NGO representative, January 2010.

7 Doing vernacularization

The encounter between global and local ideas about women's rights in Peru

Peggy Levitt, Sally Engle Merry, Rosa Alayza and Mercedes Crisóstomo Meza

- **Conceptualizing vernacularization**
- **The organizations**
- **The different types and times of vernacularization**
- **Moving forward**

How are global norms about gender actually understood and used on the ground?

Between 2003 and 2008, we studied the circulation and localization of women's human rights ideas and practices in four locations: Beijing, China; Baroda, India; Lima, Peru; and New York City, United States.[1] In each site, we compared several women's rights nongovernmental organizations (NGOs) to see if and how they translated global concepts into local terms. We asked what women's human rights look like in the day-to-day work of these organizations and sought to explain why different organizations in different contexts translated allegedly universal concepts in distinct ways.

We found that, in practice, the discourses of human rights are malleable, subject to different interpretations and meanings that activists apply to particular problems in specific situations. We call this process *vernacularization*. Instead of simply applying human rights ideas as articulated in international law and conventions to local situations, the leaders and staff in the organizations we studied redefined and adapted these ideas so that they would be easier to understand and use. They modified aspects of women's human rights so that they were comprehensible and relevant. Vernacularization on the ground is a process of creating meaning by connecting, in various ways, the discourse of the global with local social justice ideologies, within the context of a particular organizational style and ethos.

Our fieldwork revealed three types of vernacularization. The first relied on the imaginative space created by women's human rights rather than the discourse of rights itself. It drew on the aspirational possibilities created by words. Staff did not talk about rights directly in their work but used the momentum and power provided by the backdrop of global discourses about rights to advance their cause. The second type vernacularized ideas. It stretched the boundary of issues considered by women's groups by using the language of human rights to tackle new issues, thereby expanding the repertoire of what could and should be fought for. In the Indian case, staff used English words and linked them to local narratives and symbols to connect the human rights framework to the struggle for sexuality rights. They appealed to the magic and allure of the West and combined it with a strong claim that these same ideas had deep Indian roots. The third type of vernacularization involves using the core concepts of women's human rights, articulating them in locally appropriate ways, and putting them into practice. It goes beyond the realm of ideas to the domain of practice. Staff explicitly referenced women's human rights to encourage the women they worked with to shift their understandings of self and act upon them in familiar and acceptable venues like street plays and community celebrations. Like its second counterpart, this third type of vernacularization expands the range of issues organizations consider but also specifies appropriate technologies, behaviors, and arenas to fight for them.

In all three types of vernacularization, the ideas and resources women used on the ground did not rely directly on the texts of international human rights law or on United Nations (UN) declarations. They did not appeal to international bodies nor did they reference particular articles or sections of conventions. Indeed, we heard very little discussion about the Convention on the Elimination of All Forms of Discrimination Against Women (CEDAW) or the UN apparatus. These organizations use women's human rights largely as discursive and aspirational resources rather than as legal ones. In fact, a second kind of reworking and adaptation would be needed to move from women's human rights as a concept for social movement/mobilization to using rights as law. UN Women, the new body that merged four UN women's agencies, which are the United Nations Development Fund for Women (UNIFEM), the Division for the Advancement of Women (DAW), the Office of the Special Advisor on Gender Issues (OSAGI) and the International Research and Training Institute for the Advancement of Women (INSTRAW), might have a special role in advancing this process.

In this article, we focus on the case of women's rights in Peru. The Peruvian case compellingly illustrates how the pathways through which rights travel strongly influence how and where they circulate. It also drives home the ways in which vernacularization changes over time, in response to different changing geopolitical conditions.

Conceptualizing vernacularization

We understand the circulation of global rights discourses as a transnational process. We see the beneficiaries and staff of the NGOs we studied, and the organizations themselves, as embedded in transnational social fields where they are influenced by multiple sets of laws, cultural repertoires, and institutions. Their daily rhythms and activities respond not only to more than one state but also to social institutions and networks, such as religious communities, that operate within and across borders.[2]

Our research revealed several kinds of *global values assemblages*[3] that are critical to the spread of ideas about women's rights. Their emergence depends not only on the work of individual activists but also on the broader geopolitical context, influenced by events like the end of the Cold War, the decline of socialism and communism, and the rise of the Washington Consensus. A neo-liberal assemblage promoting structural adjustment, democracy, capitalism, human rights, the rule of law, transparency and accountability, and gender equality that is spread by institutions like the World Bank and the Ford Foundation is perhaps most familiar. A fundamentalist religious assemblage based on gender complementarity, tradition, conservatism, and authority, which is spread by networks like the Tablighi Jamaat or the Universal Church of the Kingdom of God (UCKG), transnational religious movements and organizations with followers around the world, is a second example. A third "anti-globalization" assemblage revolves around anti-consumerism and anti-materialism, environmentalism, fair trade, and living locally and simply.

These assemblages enable and are enabled by the circulation of a related, global feminist assemblage based upon the notion of women as autonomous subjects, capable of owning property, divorce, inheriting money and land, earning income, expressing opinions, and being protected from violence.[4] This assemblage views women as equal participants in public and private and as embracing an agentic, self-interested, secular, rational self. It embraces gender equality rather than gender complementarity. It is expressed in a set of national and international laws and practices such as CEDAW, international women's conferences, International Women's Day, and the theoretical work of the many women's and feminist studies programs that have proliferated at universities over the past 30 years.

By following the trajectory of ideas from global documents, theoretical writings, and international conference discussions through the talk of national leaders and NGO heads to discussions with staff members and clients, we can decipher the links between grassroots and global discussions about women's rights. The organizational channels and networks through which these values circulate strongly influence where they are appropriated and by whom. Moreover, in each site, these assemblages encounter local social justice ideologies, gender norms, and an organizational ecology that strongly influences how and where they land, what they combine with, and how and by whom they are appropriated. In some cases, the connections to the global are quite tenuous, while in others, the international reference and substance remains strong.

This process of appropriation and customization is called *vernacularization*. We see translation and vernacularization as slightly different processes. Translators communicate something so it can be understood while vernacularizers modify or alter it strategically so it becomes acceptable as well. Vernacularizers exercise agency, purposefully combining elements of the cultural repertoire already in place with specific pieces of what is circulating so that they are comprehensible and useable in a particular context. Vernacularization also differs from the more top-down processes of norm diffusion and implementation, as conceptualized by human rights scholars, because it involves constant interaction and exchange between the ideas, practices and technologies circulating within the transnational social fields where it takes place.

The organizations

In Lima, we studied two organizations working to improve the lives of poor, urban women.[5] They each had clear, though different, understandings of the causes, consequences, and remedies for gender inequality as it evolved in the Peruvian context. They both were embedded in international organizational networks and drew upon global discourses about women's rights but in distinct ways.

Servicios Educativos El Agustino (SEA) is a non-profit organization, founded by the Jesuits in 1978, to promote social justice and fight poverty among the urban poor. SEA provides an integrated package of services to its clients, addressing their social, economic, and political needs through strategies based on progressive Catholic social thought. SEA is not fundamentally a women's rights organization. Staff members draw upon Catholic Liberation Theology to address the spiritual, physical and social needs of those they serve. They also emphasize the importance of working collaboratively with other NGOs, Catholic organizations, and state agencies.

SEA began its work by offering job training to men. When Peru's economic crisis worsened in the 1980s, it broadened its focus to help groups of women establish *comedores populares* or communal dining rooms/canteens. Members of the *comedor* pool their money and resources to cook together at lower cost. This program has since expanded to include leadership training, domestic violence prevention, and micro-credit programs.

The comedores populares in El Agustino opened with support from Caritas and the Comisión Episcopal de Acción Social (the Episcopal Commission of Social Action, CEAS) in 1979. They also received some support from the government, through donations of food and utensils. As we outline below, this changed during the 1990s, when the Fujimori government stepped up its efforts to control grassroots activism by creating the Programa Nacional de Apoyo Alimentario (PRONAA) which offered help to comedores in exchange for their political backing.

According to Carmen Lora, an advisor to the comedores and an expert on gender:

> The *comedores* were not only spaces of cooking, they were also educational spaces where women learned what their rights were, what was going on in Peru, what caused poverty, why hunger existed, and how they could improve their situation. It was not just about lowering costs. From the very outset, it was also very much about women organizing and running these organizations on their own.

The *comedores* populares produced several kinds of leaders—those who ran the day-to-day operations of the kitchens, those who worked with other neighborhood committees, women who worked at the district level, and women who became city-wide and national-level activists. While each level worked autonomously, social ties between individual leaders connected zone-, district-, and municipal-level activities. The women who ran the day-to-day operations of the comedores were more likely to draw on local cultural frames (although small numbers occasionally participated in national workshops). The clergy and lay professionals who support SEA are more likely to travel, participate in training abroad, and to be well versed in the teachings of Liberation Theology.

In the early 1980s, there were 18 *comedores* in El Agustino. By 2007, there were 140. In some cases, their activities also included income-generating and communal banking projects. The *comedores* in El Agustino are united under the umbrella of the *Central Distrital of Comedores Populares de El Agustino* and nationally under *La Confederación Nacional de Mujeres por la Vida y el Desarrollo Integral* (CONAMOVIDES), an organization of comedores throughout the country.

Through its work in El Agustino, SEA also became aware that many of its clients confronted domestic violence. The organization mounted a multi-pronged response including violence-prevention; courses on human rights, women's rights, health, and nutrition, sensitivity training for community workers and government officials, and establishing a safe house for victims. It received support from the local parish and police station. It also collaborated with other institutions including the Ministries of Women and Health.

While SEA takes its cue from national and international progressive Catholic discourse, El Estudio para la Defensa de los Derechos de las Mujeres (DEMUS), the second organization we studied, is more firmly embedded in regional and international feminist networks. Its staff consists of psychologists and lawyers, who identify explicitly as feminists, and who believe that bringing psychology, the law, and feminism together will constitute a powerful, effective way to improve women's status.

DEMUS was founded in 1987. At various times, it has received support from the Ford Foundation, Womankind (a British aid agency), the United Nations Development Fund for Women (UNIFEM), and Hivos (a Dutch NGO). It began by offering legal assistance to victims of family and sexual violence through a telephone hotline. Some staff members feared this approach came up short. They believed the law alone could not right gender wrongs and that women also needed support and counseling. In 1994, DEMUS added psychological counseling to the array of services it offered.

The DEMUS website reads, "We want to bring to light violations of the human rights of women and to promote a social response to this serious problem. We want our work to help build a society based on gender equality and human rights."[6] In our conversations with staff, they described what led them to abandon terms like "domestic" and "family" violence and talk instead about "violence against women." They define this as "any act that has as its end the control or domination of women, based on their gender, due to the inequality of power between men and women, that affects the exercise of their rights and their full development as human beings."[7] In adopting this language, DEMUS joins with feminist activists against domestic violence in many other countries. To redress these inequalities, they currently pursue two lines of work—"Communication and Citizenship" and "Services."

The first line of work focuses on changing public opinion and influencing public debates. DEMUS does this by forming alliances with NGOs, community organizations, government agencies, and national and international agencies, including the Latin American and Caribbean Committee for the Defense of Women's Rights (CLADEM, a regional network), the Movement for Equal Opportunity Collective, and the Tribunal for the Economic, Social and Cultural Rights of Women. Their Latin American feminism differs from feminism in the United States and Europe which, in their view, privileges the white, middle-class experience. While they receive funding from international agencies like the Ford Foundation and Womenkind, staff stressed the partnership-like quality of these relationships, the mutual learning that occurs, and the complementary nature of their efforts. In fact, the coordinator of Womankind in Peru noted, "Womankind can also learn from DEMUS. The legal work they do, the investigations, the type of cases and their lobbying before the InterAmerican Court."

DEMUS staff also work as consultants at the "women's police stations" established by the government to serve women better and act as "watchdogs" in municipal courts and government offices. Peru was one of the first countries in Latin America to adopt special legislation on domestic violence. The Law for Protection from Family Violence (hereafter referred to as Family Violence Law), first adopted in 1993 and subsequently strengthened in 1997, expedited the process of dealing with domestic violence cases and clearly defined roles and responsibilities within the justice system.[8] In addition, in the 1980s, the government created 12 women's police stations to respond to domestic violence. Twenty specialized sections were also created within existing police stations. In the early 1990s, the government also established Municipal Defender's Offices (Demunas), which are now one of its first lines of defense against domestic violence. The most recent innovation has been a system of one-stop centers, established for victims of domestic violence by the Ministry for the Promotion of Women and Human Development (PROMUDEH), where women can meet with police officers, medical examiners, and state prosecutors under one roof. DEMUS provides pro-bono assistance on legal cases involving women. They work with journalists and mount public education campaigns to counteract discriminatory messages about women, particularly in the media. Finally, DEMUS runs a leadership training school for community activists and government officials. They see themselves as training citizens by teaching participants political skills and developing strong, self-actualized individuals who are physically and emotionally self-confident.

The different types and times of vernacularization

Our findings reveal examples of three types of vernacularization and the ways in which vernacularization changes over time in response to different political and social currents. The vernacularization of global ideas about women's rights in Peru took place during four key periods of encounter between the organizations promoting these ideas and the women intended to receive them. These encounters were shaped in fundamental ways by the national context, particularly massive rural to urban migration and the widespread integration of the poor into national politics during the 1970s and 1980s. The women most affected by these trends, who participated actively in civil society, came to understand that the goals they were pursuing in other arenas found a potent ally in the human rights movement.

The idea of social justice

From the outset, the circulation of rights was intimately linked to the widespread acceptance of the concept of social justice. Throughout the 1970s and 1980s, many religious organizations, feminist groups, and NGOs on the Left embraced "the fight for social justice" umbrella and spoke loudly and clearly about it to poor urban residents. In response, many of these groups altered what they did and how they did it—changes that paved the way for the successful dissemination of ideas about rights that followed. These activists built upon the momentum and power of ideology and activism surrounding social justice struggles to introduce the fight for human rights. Because the new Left drew much of its inspiration from Marxism, these activists agreed with social theories that saw Latin America as perpetually economically dependent and underdeveloped. Their activism was based on a clear class analysis and one of their primary goals was to mobilize previously marginalized groups. When ideas about women's rights were introduced, they landed on fertile ground.

Liberation Theology also inspired religious communities to redress injustice and suffering among the poor—a second assemblage of ideas that resonated with rights claims. Both religious and lay leaders embraced Pope John Paul II's call to pay special attention to the downtrodden. In the past, this commitment had translated into a focus on the individual—church workers helped individuals cope with their everyday challenges without addressing the underlying root causes of their poverty. Now, these activists articulated a structural analysis of Peruvian inequality and questioned the church's role in creating it, not only building on the political space created by these ideas but vernacularizing them such that they resonated strongly with other struggles and articulating concrete ways to put them into practice. They stressed the need to transform individuals and social institutions at the same time. The Christian belief that every individual has dignity, no matter how marginalized or discriminated, transformed the poor into legitimate social actors who had the right and responsibility to actively pursue social change.

At the same time, a feminist perspective, as articulated by its middle class proponents, was also being vernacularized. By focusing on family violence and on liberating women from their relegation to the private, domestic sphere, DEMUS staff pared down a broad global women's rights package to its most relevant components. When women participated in consciousness-raising groups, lectures, training, and workshops, they realized they were persons with rights who had the power and voice to be seen and heard.

These social actors did not work in isolation. Rather, they came together in powerful partnerships that had not been forged before. The 1960s and 1970s saw Peruvian and foreign-born priests and nuns, members of the Catholic laity, and university students who identified with the Left and embraced the "preferential option for the poor" coming together to devise agendas and strategies for social change. Vernacularization, in this case, generated new ideas and forms of pastoral work in local parishes. The neighborhood *comedores populares*, in fact, exemplified this emerging link between religiously inspired activists and the Limeño poor.

SEA, in particular, saw women as important protagonists in development. It built bridges with other social activists on the basis of their shared commitment to social justice that gradually expanded to include rights. Some of SEA's leaders went on to direct regional and national coalitions of food kitchens, applying ideas from one struggle to another, which also furthered their social agenda.

While these activities took place locally, they were strongly influenced by global factors. The questions raised by the Left and by feminists were inherently global in nature and informed by the travels and trainings attended by lay and Catholic leaders abroad. They took hold in Peru precisely because national social and political elites were connected to academics, activists and politicians around the world who were also concerned about these larger issues. Circulation occurred in multiple directions. For example, while Liberation Theology originated in Peru, it then spread throughout the world through Catholic Church and academic networks.

In short, the stage was set for vernacularizing ideas about rights through a combination of the heightened awareness about social justice among important sectors of Peruvian society and the transformation of certain key institutions such as the courts and the criminal justice system. Organizations on the Left, feminists, and religious activists each contributed to this fortuitous convergence. In response, the poor came to see themselves differently and to reinvent their relationship with their counterparts, the state, and other social and political organizations.

Facing up to family violence

The establishment of training programs around family violence, and the legal mechanisms for addressing it, marked a second stage in the vernacularization of ideas about rights that reflected the changing political times. Between the mid-1980s to the mid-1990s, NGOs, academics, the state, and international organizations like the UN, each helped elaborate and spread ideas about rights. A host of training courses, workshops, and seminars on domestic violence and women's rights for poor urban women helped to make these ideas palatable to local women and introduced strategies to achieve them.

That so many activists were influenced by a "gendered" perspective and that Women's Studies gained ground as a discrete discipline also drove vernacularization during this period. Several aid agencies took on gender equality as their focus. By calling patriarchy into question, women realized that they had the right to live without fear, a personal dimension that had not been clear before. They began to question their experiences of domestic violence and to help others come to grips with their own experiences of abuse. Not all women became collective activists. But by participating actively, some became stronger and more mature leaders in the day-to-day affairs of their communities and in the social movement as a whole.

The testimonies we collected revealed that vernacularization posed difficult challenges. While some women felt more affirmed and autonomous as individuals, others spoke of subsequent familial conflicts and ruptures. The ideological climate had shifted but the social and economic structures undergirding it had not. All of these experiences, however, contributed to the creation of citizens: When a woman recognizes that she has the right to an opinion, when she ventures outside of her home environment, and when she becomes aware of her own self-worth, she begins to imagine a new type of life not characterized by violence and discrimination, which she has the skills to shape actively. She embraces a new, autonomous subjectivity which is worthy and capable of exercising rights.

Organized women: between violence and manipulation

The third historical moment of encounter between ideas about rights and their potential recipients occurred between 1990 and 1998. Widespread political violence plagued Lima's poor districts. The Fujimori government intervened actively in the day-to-day operations of many NGOs, because it saw this as a way to gain access to, and ultimately manipulate, potential political opponents. The language of rights was co-opted to pursue the state's agenda.

How could women's rights ideas be disseminated in this difficult context? Members of Sendero Luminoso, a Maoist-inspired terrorist movement which waged a protracted civil war to bring about radical, fundamental social change, considered everyone working in democratic politics questionable, if not threatening, because their work "lessened the contradictions" that would engender revolutionary struggle. Because Sendero wanted to worsen poverty and hunger to spark a fuse, they threatened and intimidated women working in the *comedores.* In contrast, the Fujimori government positioned itself as defending human rights.

NGO and parish workers tried to help activists fight Sendero and to deflect political manipulation by the government. They offered psychological support and asylum to leaders who were threatened. But inevitably, the environment of intense violence and fear altered the partnerships between women's groups, NGOs, and the people they tried to serve. Because these diverse actors could no longer communicate with one another as freely and as often, they grew distant. Little vernacularization of any type occurred.

In fact, it was the Fujimori government that posed the biggest challenge during this period because of the "deal with the devil" it offered to NGOs. Through incentives, and sometimes through barely veiled force, the government actively encouraged women's groups to embrace its policies. To receive programmatic aid or material resources, NGOs had to sign on to the government's program. Groups that allied themselves with the government were rewarded resources with which to develop reproductive health and birth control programs. They were later strongly criticized for promoting involuntary contraceptive surgery or for endangering rural women who were operated on under precarious conditions and without informed consent.

Other groups, such as DEMUS, challenged the state's dictatorial approach and refused to become its bedfellow. It took courage to take such a stand because, during this time, local authorities were routinely harassed, and even sometimes killed, by the government.

Little by little, the leadership of the neighborhood food kitchens changed hands. Some of the original leaders acquired enough money and skills to move up and out. Others left for more negative reasons. Even though cooking together still lowered expenses, fear destroyed the groups' spirit and solidarity. Distrust and indifference increased; there were complaints about members not doing their fair share. Moreover, younger women with more direct links to the government assumed positions of power and tried to influence leaders who were poorer, less experienced, and more vulnerable. Many believed the government donated food to the kitchens in exchange for support. During this period, women's organizations fell into two camps—those allowing the government to control its activities in exchange for resources and those who insisted on maintaining their independence.

The concept of women's rights

Explicit discussions about women's and/or human rights marked the fourth period of vernacularization, which took place during the late 1990s, when widespread public protests demanded regime change. This period is a time when the fruits of all three types of vernacularization reached fruition. Broad sectors of society, including young people and women, marched through the streets protesting the dictatorship. They demanded that their rights be protected by the state, that the rule of law be reinstated, and that civil and political rights be respected, especially in the regions where armed conflict had occurred. Different social groups, such as the Leadership School in San Juan de Lurigancho and the Network of Legal Consultants in El Agustino, once again came together. Participants were groups of women connected to NGOs and parishes who identified with ideas about empowerment and women's rights. They refused to be manipulated by the state.

Towards the end of the dictatorship, as the demands for democracy grew, so did the demand for equality and transparency in the social and political process. This led women's groups to act against the government which, in turn, laid the foundation for their mobilization during the democratic transition.

In sum, a women's rights global values assemblage circulated widely in Peru where it was vernacularized to fit with national and local approaches to social justice and was adopted by local women in different forms during different periods. In Lima, women's rights travel over terrains marked by years of government-supported urban collective action begun under the Velasco regime during the 1960s. They resonate with Andean traditions of collective labor sharing, governance, and land tenure that rural peasants brought with them during the great internal migration to Lima during the 1950s and 60s. Human rights activists mobilized when Sendero terrorized rural peasants in the nation's highlands and when the Fujimori government abrogated rights to crush them. The government's Truth and Reconciliation Commission, put in place to heal the nation, also employed a rights vocabulary. Activists drew on ideas disseminated at international women's conferences and through international resolutions and conventions including the 1993 Vienna Conference on Human Rights and the Convention of Belem do Para in 1994.

Our case reveals that this rights assemblage traverses at least two pathways. In the case of SEA, global ideas about women's rights combined with global and national ideas about Catholic social justice. Some factions within the Peruvian church listened carefully to Pope John Paul II's call to serve the poor. Gustavo Gutierrez, one of the creators of Liberation Theology, was strongly influenced by Vatican II. The poor became the central focus of the Church's pastoral work because they were seen as messengers of the Gospel. Church leaders, and the lay activists working with them, who saw themselves as putting Liberation Theology into practice, urged the poor to face up to the historical conditions preventing them from living with dignity. They drew on familiar models of community organizing, including the creation of comunidades eclesiales de base (Christian-based communities or small groups of social support and prayer) and Andean collective labor-sharing traditions. They did so in conversation with similarly inspired activists across the continent and across the world. Thus, though SEA is locally grown, it is deeply embedded in strong national and international religious networks.

DEMUS, in contrast, is a national NGO based in Lima with strong linkages throughout Latin America. It forms part of the Peruvian women's movement, with its well-established history and significant support from US and European NGOs and foundations. Its leaders participate regularly in regional and international feminist networks. It draws upon a particular tradition and version of feminism in Latin America, often positioned in opposition to feminism in the United States which is considered to be dominated by white, middle-class interests and out of touch with the concerns of poor, rural, and indigenous women.

Both SEA and DEMUS talk about rights but the way they understand them and how they want to pursue them differs radically. Their conceptions are also very different than those depicted in CEDAW. Not surprisingly, when SEA staff vernacularize, they draw upon and reshape the religious layers of the Peruvian cultural terrain. Women have rights because they are children of God living in the Kingdom of Christ not because they are individuals endowed with rights because they are human. A woman has social and economic rights and the right to be free from hunger and violence. SEA pursues these goals through well-worn Catholic social mobilization strategies—the Christian-based community, a discourse of compassion, morality, and collective responsibility, and a clear authority structure.

DEMUS also promotes freedom from violence—however in contrast to SEA—in conjunction with the right to be a free, self-actualized, autonomous subject in charge of her own sexuality. In this case, vernacularization involves stressing the connection to international feminism by focusing on sexual and reproductive rights. It sees violence against women and gender inequality as the product of power inequities between men and women in the public and private sphere. Redressing them requires working through the legal system and dealing with its psychological and socio-cultural roots. Thus, DEMUS aligns itself with groups that see themselves as aiding rights-bearing individuals who achieve self-actualization through the self-help, therapeutic culture of the industrialized west. It also draws upon the rhetoric of democratization, equal representation, and social mobilization that is widespread among progressive social movements.

The adoption of the concept of rights by poor urban women in Peru resulted, in part, from their contacts with NGOs, parishes and other organizations which disseminated these concepts. But this is not the whole story. Ideas about rights landed on soil that was fertile in different ways during the four periods we outlined. Poor men and women mobilizing for the first time brought about dramatic national changes. The rules of the game in Peruvian society have slowly and steadily changed, including a shift from an exclusive oligarchic society to an incipient democracy in which greater numbers exercise citizenship and the Peruvian state acknowledges the diversity of its constituents. Because these new groups expanded the notion of citizenship, the state had to respond in new ways. Women also became social and political protagonists, a transformation that would not have been possible if large numbers had not incorporated ideas about rights and citizenship into their understandings of self. As women invented and articulated these new subjectivities, they became more open to new ideas, philosophies, and approaches, which also transformed how they interpreted and appropriated the new information they encountered.

Moving forward

What does this mean for our understanding of the transnational transfer of ideas and practices such as those of human rights? Global values assemblages circulate widely but are appropriated differently, depending on the cultural and historical context and the pathways and terrains over which they travel. Vernacularization matches global ideas about women's rights with their national and local counterparts. In our larger study, this included social work ideology in China, Gandhian thought and socialism in India, as well as Liberation Theology in Peru. The women's rights assemblage was also connected to different organizational forms and technologies in each context including communal kitchens and labor-sharing practices and Catholic base communities in Peru, government Women's Federation and universities in China, caste and village panchayats in India, and civil and minority rights activism in the United States. NGOs and activists in each site piece together distinct collections of discourses, practices, and technologies of activism drawing in various ways on the human rights framework.

Vernacularization is a widespread practice that takes different forms in different kinds of organizations and, by so doing, brings human rights as a justice ideology into a wide range of communities. At the same time, it produces a set of ideas and practices that are remote from the original documents, such as CEDAW, and that often differ from the ideas articulated by international organizations and agreements. While many of these documents envision new kinds of laws and courts granting and guaranteeing rights supranationally, in all of our cases, most activists on the ground embraced rights as an ideology and social movement. Vernacularization fueled the ideas and attitudes, and people's mobilization around them, but this did not translate further into creating international legal and criminal justice institutions that are the goal of many conventions.

This belies a disparity between those who see human rights as law and those who mobilize these ideas for social movements. This is a significant but largely unrecognized difference. These are different actors with different training. The first group are largely lawyers, the second social activists. The former group focuses on building cases and appealing to UN agencies while the latter works on building public consciousness and mobilizing popular support to confront governments. Human rights law is embedded in documents, interpretations, and processes of monitoring, while human rights social movements draw on the legality of human rights in a broader and more open way, focusing on how these ideas are adopted and transformed. These groups do not always see eye-to-eye nor do they want the same things. Activists and UN Women should keep in mind these differences when supporting the implementation of women's human rights on the ground.

Clearly, human rights law, particularly that part that emphasizes civil and political rights, is quite compatible with neoliberalism and privatization, with its focus on choice, autonomy, and individual agency. When social movements seize human rights ideas, however, they escape those textual constraints and become something else, more responsive to other sets of ideas about gender and justice. There is, in other words, a tension between human rights law as a set of doctrines and institutions for monitoring and pressuring governments and human rights as an idea mobilized in social movements. The two need each other, but they are also fundamentally different and often incompatible. Human rights social movements make claims that are not necessarily in the law. They expand the domain of human rights by coming up with new rights and defining new issues. They use human rights rhetoric to generate public support and produce outrage, seeking to inflame the public to support their cause. The legal dimension of human rights legitimates their claims and provides an appearance of state power. On the other hand, human rights as law needs social movements. The implementation of human rights law depends on publicity, on public outrage, and on pressure and shaming of recalcitrant governments. This is a symbiotic but tense relationship.

Notes

1 We carried out this research in collaboration with Liu Meng at the Department of Social Work, National Women's University in Beijing, N. Rajaram at the Department of Sociology, Maharaja Sayajirao University in Baroda, India, and M. Rosa Alayza at the Escuela de Graduados, Pontífica Universidad Católica del Perú in Lima, Peru. The results of this research are outlined in a special volume of *Global Networks* 9, no. 4 (2009).
2 Peggy Levitt and Nina Glick Schiller, "Transnational Perspectives on Migration: Conceptualizing Simultaneity," *International Migration Review* 38, no. 3 (2004): 1,002–1,040.
3 These may be loosely or tightly clustered, created from actors, objects, and ideas traveling at different rates and rhythms, across the different scales and scopes of the social fields within which they takes shape. Gilles Deleuze and Felix Guattari, *A Thousand Plateaus: Capitalism and Schizophrenia* (Minneapolis, MN: University of Minnesota Press, 1987).
4 Amrita Basu, with the assistance of C. Elizabeth McGrory, ed., *The Challenge of Local Feminisms: Women's Movements in Global Perspective* (Boulder, CO: Westview Press, 1995).
5 We chose to study these organizations because they represented two very different sites of women's rights activism and are connected to and receiving "material" from very different transnational networks. One is a Catholic, anti-poverty organization connected to other religious organizations around the world while the second is a feminist organization that relates to global women's rights groups, particularly in Latin America.
6 DEMUS, *Informe sobre el Servicio Integral de Orientación Telefónica* (Lima, Peru, 2003).
7 DEMUS, 2005 internal document.
8 Human Rights Watch 2000, www.hrw.org/backgrounder/wrd/peru-women.

8 Translating international norms

Filters to combating violence against women in Lebanon

Rita A. Sabat

- **Translating violence against women: a framework**
- **The Lebanese state, violence against women, and the campaigns to eliminate violence against women**
- **Filters of translation: culture of violence, family honor, and family cohesion**
- **Conclusion**

The Beijing Platform for Action (1995), the Declaration to Eliminate Violence Against Women (1993), and the Convention on the Elimination of All Forms of Discrimination Against Women (CEDAW) (1979) have established international norms for combating violence against women. Countries around the world are applying these international norms within very different national contexts. Lebanon, a country reeling from civil unrest and characterized by a complex socio-political system based on confessional divisions, is one of the countries that has taken up the battle. This paper traces what happens to an international norm as it transcends boundaries to be enacted in a local environment. Is this norm simply transposed and faithfully implemented into local laws and society? Or is it translated and filtered in a way that changes the norm? What are the mechanisms at play in the process of implementation and what social filters operate in the translation of norms?

I trace the implementation of the international norm by exploring Lebanese campaigns on violence against women. CEDAW does not provide an explicit definition for violence against women; however, general recommendation 19 put forth by the Committee on the Elimination of Discrimination Against Women has recognized that "gender based violence against women is violence that is directed against a woman because she is a woman, or violence that affects women disproportionately. It includes acts that inflict physical, mental or sexual harm or suffering, threats of such acts, coercion and other deprivations of liberty."[1]

Accordingly the committee agrees this definition is tacitly covered under article 1 CEDAW. In Lebanon, violence against women is found within the purview of the state, in society, and in civil confessional legal systems.

Most activism targets the issue of violence against women within the context of family. This is a particularly burning issue because of the communal social organization of Lebanon and because of the absence of legal protection for women within the family. There are no accurate statistics on domestic violence; however, research indicates that the problem is widespread, and estimates suggest that up to 30 percent of homes have experienced some form of domestic violence.[2]

To understand violence against women in Lebanon, it is important to underline the experience of violence at a wide scale through 16 years of civil war (1975–90). While the war took place prior to the emergence of a global gender equality regime that identified the problem of violence against women, women's organizations in Lebanon can be traced back to independence in 1943. Lebanon's activity on this front was limited in scope to a few social organizations and researchers. Dar Al Amal Association founded in 1970 is an example of an organization that addressed a form of violence against women for years prior to the international movements' push in the mid-to-late 1990s. The programs of this organization targeted vulnerable women and children and offered rehabilitation and prevention programs within the scope of a civil war.

This article traces how women's organizations campaigning on violence against women have evolved in Lebanon. The research involved individual, open-ended, taped interviews with lead questions conducted in Arabic, French, or English with members of organizations, leaders, academics, activists, and politicians in Lebanon, all of whom have worked in the three campaigns analyzed. These interviews shed light on the process of translation of international women's rights norms at the local level. Interviews helped reveal how organizations collaborate and at times clash to achieve equal socio-political rights while navigating a messy and confusing geopolitical atmosphere. More poignantly, the open-ended interviews allowed for an honest engagement with women's rights in a multi-confessional setting thereby revealing the expectations and commitments regarding the international gender equality norms. Lebanon's multi-confessional role is important in this context because all issues pertaining to women are housed within the confessional communities' laws leaving women across confessions unequal to each other. However, in the context of violence against women there are many ways the multi-confessional character of the state will not be visible in the contours of this work. The interviews also revealed resistance to these norms situated in complex geo-political contexts.

The first section of this paper introduces the framework I apply to illustrate how international norms are transposed. That is followed by a brief background on Lebanon and the conditions of violence against women including some brief information on the existing campaigns and social conditions. The third section illustrates how the international norm on combating violence against women has undergone a transformation through the campaigns examined. I identify a mechanism involved in the process referred to as "filters of translation." The two following sections lay out how this mechanism becomes operational in the campaigns examined. In the conclusion, I will review the process of norm application, how it has changed the norm and what this could mean for future work.

Translating violence against women: a framework

Literature in International Relations and on social movements has developed models that trace the implementation of norms assuming the norms are static in nature and do not change in the process.[3] They imagine that the end stage of implementation is domestic norm internalization.[4] This literature largely neglects the role of domestic actors and structures and their impact on norm translation. In contrast, I argue that the local adoption of a norm entails a metamorphosis of the norm resulting from domestic filters of translation and from the intervention of domestic actors. I seek to widen arguments about norm implementation drawing on more recent critical constructivist work of scholars such as Jeffrey Checkel, Andrew Cortell and James Davis Jr., and Susanne Zwingel who have called attention to the domestic factors that impact norm implementation.[5] For international norms to become a matter of practice at the local level, they do not simply get implemented or diffused, but they go through a process of translation. Cortell and Davis point in their model to the way international norms are introduced and embedded in the domestic structure of the state. One of the factors they suggest as relevant in the translation of international norms is "cultural match"—that is the extent to which national and international understandings coincide. Similarly, Checkel stresses that various domestic factors influence the level of legitimacy that an international norm can acquire at implementation, while Zwingel analyzes the processes and actors involved in the implementation of gender equality norms via CEDAW. I am interested in understanding the impact an international norm may have in varying local contexts, but also want to show how a norm changes in the process of implementation.

Following Checkel, Zwingel, Cortell and Davis, and contrary to most existing theorizations of norm implementation, I suggest a more process-oriented understanding of norm diffusion, one that involves "filters of translation" which show the changes that happen to an international norm during its transition phases. Translation includes the literal language translation but also a translation that involves dialogues around the issue that bring it in line with domestic sensibilities. Susan Bassnett and Harish Trivedi define translation as something that does not just take place in solitary confinement, undisturbed, and without layers of complexity: "Translation is a highly manipulative activity that involves all kinds of stages in that process of transfer across linguistic and cultural boundaries."[6] Ana Lowenhaupt Tsing adds to this postcolonial notion of translation her concept of a "faithless translation"[7] that grasps translation as a "rewriting of a text in which new meanings are always forged by the interaction of languages."[8] She understands translation as an ongoing negotiation process which spawns new meanings. This process, however, is not straightforward or systematic. In a nutshell, according to Tsing "there are no originals, but only a heterogeneous continuum of translations, a continual process of rewriting in which meaning—as well as claims of originality and purity—are made."[9] I use this conception with a broadened understanding that extends beyond language to include various structures and the interests and identities of those involved in the process of norm translation. Feminist agents engage in an initial "faithless translation" because they have their own identities and interests that are deeply intertwined with identity and background. However, this translation is not final and is continuously revisited as feminist agents are confronted by arguments reflecting international, state, and social structures. I argue that norms go through the process of translation; *how* these norms change is determined by specific "filters of translation." I define "filters of translation" as context-specific repertoires of interpretation that facilitate certain constellations of arguments. These context-specific repertoires of interpretation are informed by structures as well as by interests and identities of the agents. They bring together elements of domestic social and political structure with an international discourse and the arguments and framings of political agents, including feminist nongovernmental organizations (NGOs). Filters of translation do not imply that an international norm cannot be implemented at the national level, but that what takes place in the process of implementation is not a fixed and top-down process of transfer. The "filters of translation" identified in the campaigns reflect the change in the international norms as they go through the process of translation in transition to the domestic context. In this sense they are like photographic filters, an external object that can control the diffusion of the norm, sometimes changing it subtly while at other times manipulating its entire outcome.

Relevant structures and agent identities

Agents are embedded in structures, and structures constrain what they are able to do and how they argue. Accordingly, structures influence the kinds of filters of translation that agents produce as they argue for the implementation of gender equality norms. The structures identified in the campaigns examined include state, social and international structures. I will discuss each of these in more detail.

The state structure, which includes government legislative and executive branches, also consists of the confessional system that characterizes Lebanon's institutions of government and suffuses Lebanese policy-making. Lebanon is a parliamentary republic, with a unicameral National Assembly composed of 128 deputy members where the people elect the cabinet according to proportional confession every four years, with the cabinet split 50–50 between Muslims and Christians. Another key element of the state structure is the legal system. All attempts at legal reform directly related to crime that occurs within the public sphere fall within the purview of the Lebanese Penal Code and the legislative branch. Lebanon's penal code is inherited primarily from the period of the French Mandate prior to independence; also included in the existing legislation is the legacy of the Ottoman Empire, which ruled the area for over three centuries. The foundations of family law in Lebanon can be traced back to the Ottoman Law of Family Rights of 1917.[10] At the same time, the Ottoman system of governance delegated application of family law to the confessional groups. As a result, in contract with the state, family law is governed by up to 15 systems of confessional law.

Social structures refer to the way Lebanese society is ordered. For the purposes of my inquiry I am particularly interested in the rules that outline how familial and communal life is ordered. While there are differences across confessional and regional communities in Lebanon, I postulate an overarching common social structure that is rooted in confessionalism and a kinship patriarchal model. Similarly to most Western societies, Lebanon's ordering goes beyond a basic patriarchal structure to include hierarchies embedded in kin, age, and class that operate simultaneously.[11]

Lebanon's social structure is steeped in an inherited kinship patriarchal-based communal system where gendered rules are defined within and around a family, community, and nation which is crucial for the translation of women's rights norms. In Lebanon, rights and responsibilities are earned through the web of relationships developed and hierarchically ordered according to age, gender, and the privileging of males on the basis of kinship. As individuals within the family grow older they are afforded more privileges than the younger; however, when one factors in their gender it outweighs the age factor. In summary male elders in the family are the first to access and delegate rights and responsibilities.[12] Even within the use of language, there is evidence of these underlying gender inequalities as will be examined in the campaigns. The social structure activates the constellation of arguments produced. Agents translating women's rights norms are part of the social structure therein being coerced at times by their own subject position. Translating international gender equality norms through social structures can metamorphose the norm at hand.

In the realist tradition of International Relations, the international structure is characterized by the relative distribution of state capabilities. Samuel Huntington has suggested that this structure changed after the Cold War into one of opposing or "clashing" civilizations. Arguably, the politics of the "war on terror" have contributed to bringing notions of cultural difference, including religious difference, to the foreground of international politics. Tied up into these politics are suggestions of a fundamental difference between Islam and the West, in particular with regard to understandings of women's rights. This dichotomous discourse did not appear in a vacuum and cannot be isolated from its historical and political origins within the domains of colonialism and the "othering" that has accompanied it. International norms and their adaptation to local contexts are not isolated from this political discourse and do not appear outside of the geopolitical history that gave birth to them. Such a projection of the human rights discourse has resulted in the wariness of those on the receiving end, and many have begun to perceive it to be an imperialist venture.[13] This is the international structure referred to in this paper where questions of hegemony in human rights standards and the imbalance that is produced are contested by the very agents translating the international norm combating violence against women.

In sum, the state upholds multi-confessional patriarchal laws, social structures reiterate a patriarchal kinship system, and the geopolitical context places Lebanon among the violators of women's rights—a notion that is rejected as neo-imperialist from within the country. All these structures become visible in the campaign to combat violence against women.

Efforts at implementation result in translations; structures inform arguments that congeal into filters of translation. The social and international structures become operative when the campaign seeks to produce grass-roots awareness and engages in advocacy aimed at reconstructing social norms. Structures inform the arguments of agents, producing filters of translation. This whole process cannot be viable without the active participation of agents at the frontlines demanding for the implementation of international norms. By agents I refer to activists, academics, and state representatives. They play a key role in translating international norms, but also put forward arguments that define intervening filters of translation. They are able to break through solidified structures by reconstructing the norm and providing a translation that is acceptable. They are able to travel back and forth between the different domestic and international contexts re-negotiating the norm and thus contributing to the construction of diverse filters of translation. However, these agents are driven by their multiple subject positions which are imbued with their identity, interests, and background. In such cases "filters of translation" are what shape the "faithless translations" and what feminist agents help produce. At times the feminist agents identify the "filters of translation" and try to push the international norms past them, but there are times when they succumb to what is inherently part of them, as they are products of their environment.

The Lebanese state, violence against women, and the campaigns to eliminate violence against women

The following section is an overview of the implementation of gender equality norms in Lebanon by the state's signing of conventions and women's organizations' participation in the global women's movement.

Lebanon acceded to CEDAW on 16 April 1997 with reservations that some describe as against the spirit of the convention. It also has not signed the Optional Protocol to CEDAW (A/RES/54/4) passed in October of 1999 with 102 parties to date. The National Commission on Lebanese Women, established by decree 13 of the Council of Ministers to follow up on the issues outlined by the Beijing Platform for Action, and the Ministry of Social Affairs have been proactive in the battle against violence by developing and collaborating in the campaigns launched by women's organizations in Lebanon. The state of Lebanon out of its duty to respond to international agreements and criticism has participated in some of the NGO-initiated domestic campaigns to eliminate violence against women; however, it has not really gone against the grain of a patriarchal social structure that is cemented by its law. This section illustrates just how that works by describing the campaigns to eliminate violence against women in Lebanon.

In 1995, a public hearing on violence against women in Beirut, the first of its kind in the Arab world, marked the beginning of the campaign to eliminate violence against women in Lebanon. This hearing led to the establishment of the Arab Women's Court whose initial coordinator was in Morocco and has now been transferred to Lebanon. The collaborators of the Arab Women's Court in Lebanon are key players in the launching of the local campaigns examined in this section. These campaigns have two facets, one that targets what has previously been identified as direct violence and another that targets indirect violence against women. The campaigns encompass a range of activities, including vigils, awareness seminars, legal counseling, psychological counseling, and conferences. The campaigns are a product of collaboration between state representatives or institutions, NGOs, and international organizations. They provide explicit examples of how the international norm to combat violence against women is translated and how that translation sometimes produces a metamorphic change.

In the following, I give a brief overview of three campaigns that have specifically dealt with violence against women with details regarding the participants, the events, and the life span of the campaign. The three campaigns are the national campaign for the elimination of discrimination against women in the penal code, the campaign towards the protection of women from family violence, and capacity building on the prevention of domestic violence.

The national campaign for the elimination of discrimination against women in the penal code was organized by the Lebanese Women's Network (LWN), a network made up of 14 national NGOs, in 2004 and came to a halt with political crises and instability in Lebanon in 2007. This campaign was funded by the domestic partner organizations and the campaign's purpose was to amend the gender-biased laws in the penal code. The laws that have been identified as discriminatory include articles in the penal code on adultery, rape, abuse, and impudence, and crimes of honor.[14] The purpose of the campaign was to amend the laws within a context where collaboration would not be mired in religious convictions of participants and community leaders that could affect the agreed upon task.

KAFA: Enough Violence and Exploitation, one of the main NGOs combating violence against women, is the principal organizer of the second campaign, the campaign towards the protection of women from family violence. The campaign towards the protection of women from family violence was funded by the European Commission, Mama Cash, and United Nations Development Fund for Women (UNIFEM) Jordan. KAFA works in collaboration with state representatives, NGOs, and experts on translating the international norm of violence against women through legislative action. The plan is not only to change government institutions and departments, but also to re-construct the normative understanding of representatives of government and civil society and encourage their contribution to combating violence against women. The campaign was established following a regional meeting that involved representatives from 11 Arab countries, national, and international NGOs. The meeting, "Legalizing the Protection against Family Violence" was organized by KAFA and held on 22–23 July 2006. There is no end date established for this campaign as long as all the goals have not been met. On 7 April 2010 the Council of Ministers approved the draft law and has forwarded it to the Parliament where it awaits ratification. The objectives laid out by the campaign include getting the state to pass a family violence draft law, advocacy of the issue of family violence as a human rights violation, raising awareness of the issue of family violence and its manifestations, and garnering momentum across civil society in order to successfully pass the proposed legislation.

The third campaign, capacity building on the prevention of domestic violence, designed by the Lebanese Council to Resist Violence Against Women (LCRVAW) and funded by the European Commission, United Nations Population Fund (UNFPA), and the Ministry of Social Affairs, was directed at preventing family violence (including violence against spouses, children, and the elderly). The campaign was launched in 2004 and concluded in 2005 and the main target was the state-society structure. The goal was to arm city social workers with an understanding of domestic violence and to make them aware of the common misconceptions around the issue. The campaign meant to inform social workers of the role they are to play and their responsibility in handling cases of domestic violence. This campaign included several projects, one of which was training sessions for social workers on how to identify and deal with domestic violence. The campaign was also supposed to serve as a catalyst for change by proposing solutions and alternatives. On the social scale it held various town awareness-raising sessions with a group of families where conceptions were reviewed and participants were asked what those conceptions meant to them. These meetings helped bring out the private matter of domestic violence which is usually left behind closed doors into the community. The final publication, a booklet produced in 2005, included an overview of the projects and the definitions and outcomes of family violence with all its misconceptions and consequences.[15]

How have these campaigns translated the international norm to combat violence against women in Lebanon's complex socio-political context?

The following section probes the filters of translation identified as operable during the campaigns.

Filters of translation: culture of violence, family honor, and family cohesion

Culture of violence

The social and family structures in Lebanon make it extremely challenging to deal with the issue of violence against women. A primary concern, often omitted in discussions of violence against women in other contexts, is the culture of violence developed in the course of a lengthy civil war. My last interview took place in the home of a noted lawyer and activist while at the same time the road to the airport was blocked off by burning tires. Hamra and other parts of Beirut were occupied by opposing political forces that were battling it out with guns and other weapons. The television was on and we were monitoring the situation closely; our fear was that we were heading into another civil war, and yet my interview partner graciously agreed to be distracted by my interview on women's rights. However, this discussion could not stray very far from the matter at hand. She had begun to partake in an NGO assembly on drafting law on family violence with KAFA, but decided to remove herself as a member. This is the explanation she offered for her withdrawal:

> Let me tell you what the real story is with KAFA and reasons beside the travel issue that made me decline working with them. Because they started to talk about the man and how they should do this to the man and what not, but my sister the question of violence belongs to the community. Today when people are killing each other in the street and nobody is bothered by this and that this could possibly continue and nobody cares … This society is educating violence, and the education is for society as a whole, you cannot solely hold one part of the society responsible for it … And you cannot research it unless the society is united and in agreement. You cannot campaign and talk about violence in a warring society, and talk about holding the man responsible. … Because the understanding of violence in an area of conflict is one thing, and violence in a stable and united area is another.[16]

To this activist the importance of addressing the broader implications and effects of strife were lost in the translations of violence against women in the draft law. Somehow filtering this combat against violence had to address two entangled contexts: conflict and culture. The threat of the communal clashes escalating into a renewed civil war distracts from combating violence against women. Having to confront the different types of violence occurring at the local level forces organizations to prioritize combating violence based on a hierarchy of issues.

Family honor

One of my interviewees asked rhetorically "Why is it that a man's honor is carried by a woman? And what about the woman's honor; why is man not caretaker of hers?"[17] This idea is pervasive in both Lebanese families and the communities. When I was a young teenager in Lebanon my mother, father, and aunt emphasized the importance of what I considered appropriate rules of engagement; "if you are walking on the street be careful who you stop and talk to, and if someone honks their horn at you make sure not to turn and look. Your behavior can shame the family. You know how things work over here." The most dominant concern is the one addressed by most activists and academics, and that is the social and traditional context usually referred to in Arabic as "the mentality of the society (*al zuhniyeh*)."

This mentality serves as a powerful filter of translation in the campaigns on violence against women, and in particular in those attacking "honor crimes." Continuing to propagate double standards of behavior for men's and women's sexuality, whether explicitly or tacitly, is a form of condoning violence against women. For one of my interviewees at LCRVAW a major concern was that society views honor crimes as normal because it identifies the male role as one of protector and preserver of a woman's honor.[18] Indeed, the state supports the society in this interpretation and considers honor crimes as "an outcome of a complex triangular interaction"[19] between crime, state, and society that contributes to the reproduction of gender relations in the Middle East. According to Abu-Odeh, the relation between the state and society is governed by the codification of honor crimes and is used by the state as a way to both "contain the practice of crime, whilst attempting to co-opt the emergence of new subversive sexual types."[20] Through its laws the state is able to deter murder. But by allowing for lenient punishment in certain cases of honor crimes, it helps control the behavior of women. Society also takes advantage of the loophole created by Article 562, referred to as the honor killing article, by supporting and justifying male control over women.[21] This filter of translation became evident in the national campaign for the elimination of discrimination against women in the penal code, when the agents in the Lebanese Women's Network faced a crisis that led to the split in the campaign.

According to a prominent activist and founding member of the Lebanese Women's Network, the failure of the national campaign for the elimination of discrimination against women in the penal code played out in the campaign's relationship with law makers. At one of the debate sessions, the law makers indicated a willingness to amend Article 562 commonly referred to as the "honor" article. They offered to consider a request to change the article so that it allows for leniency not only towards men who have committed murder in circumstances that would define the murder as an honor crime, but also towards women. This way they were still holding on to the power over the citizen's sexuality. But, according to my interviewee:

> We don't want that kind of change; we want the article to be completely removed. Anyways there is no such thing as a woman's honor crime; it is unlikely that a woman is going to kill her husband if she catches him with another woman. It is the woman that is considered the man's honor, not the man that is a reflection of her honor.[22]

But the other partner organizations who participated at the session agreed to the proposal which also reflects the hesitancy to relinquish deep-rooted traditional positions. The result was a split in the campaign between those who wanted Article 562 to be removed entirely and those who wanted to play it safe and ask for an amendment to the article. In the end, those who supported an amendment withdrew from the network and the campaign, and with their support parliament amended the law to allow for lighter sentences for male as well as female honor murders. This kept in place the justification of honor killings; however, it now applied equally to men and women.[23]

My interviews with women's organizations highlighted how language is used to structure gender relations and reproduce the notion of family honor. At KAFA headquarters the conversation began at the receptionist desk as I awaited my interviewee. I mentioned how difficult it was to translate the conception of women's rights from English to Arabic. Her response intrigued me as she made it known that it was not the first time that this problem was broached in the office. So as my interview began I felt the need to raise the topic again, and immediately without hesitation my Arabic translation of the word woman was corrected.

> Well we don't use the colloquial term for women, we use the standard Arabic pronunciation because when the former term is used it

> can have a negative connotation, as if you are not a virgin. They always differentiate between girl and woman, and if you call a young unmarried woman, woman using the colloquial connotation then it is an insult. We always try to get people to use the common standard Arabic because used as such Woman's Right and Obligation resonates more.[24]

Colloquial language reflects underlying gendered inequalities. The translation of gender equality norms runs up against such translations. In the local colloquial context the common term for man is not accompanied by any negative or positive social understanding. However, for women language serves as a constant reminder of the value of their virginity and how it is attached to their honor and the honor of their families. Feminist agents here try to translate international norms of gender equality by literally also translating "woman," i.e., by using the formal standard Arabic language. That choice allows them to diffuse the local negative connotations attached to the word, and secondly to convey the seriousness of the issue of women's rights. Feminist agents' translation of woman helped them circumvent deeply embedded understandings on the streets and in colloquial representation. This type of translation does not conform to the implementation of an international gender equality norm but rather shows how campaigners found it necessary to be co-opted to the social value placed on woman's honor found in discussions of virginity.

Family cohesion

The other filter of translation operative in the violence against women campaigns is family cohesion. As wives and mothers, Lebanese women are perceived as ultimately responsible for the preservation of the family. Because women are not only considered carriers of honor but protectors of the home, a lot of pressure is placed on them to appropriately manage the household and the relationships within it. This leaves women with a great sense of guilt if they are unable to live up to that ideal. In cases of domestic violence women allegedly violate the ideal. Women who are caught in this web of interpretation are less likely to stand up against such violence, more likely to blame themselves for the violence and fear familial repercussions if they were to make such violence a public matter.[25]

In my interviews it became painstakingly clear that translating the international norm of combating domestic violence came up against a tough filter of translation process centered on the value of family cohesion. The problem lay in what to do if you are a Lebanese woman in a violent familial relationship. There is broad agreement among experts internationally that there is no reasoning with an abusive husband. The welfare or therapeutic models, which involve dialogue or engagement with the abuser in order to preserve the family entity, used as solutions to family violence were studied and dismissed as inappropriate remedies during the 1986 expert group meetings on domestic violence in Vienna.[26] However, both the Ministry of Social Affairs in 2005 and my interviews indicate that this is still a predominant remedy advocated in Lebanon.

As in society, so within the state there is a strong commitment to preserving the family at all cost as evident in a report issued by the Women's Affairs Section of the Ministry of Social Affairs, on 19 September 2005.[27] In a response to the Special Rapporteur on Violence Against Women, the Ministry of Social Affairs Women's Section wrote that their work on combating violence against women within the state structure included a constructive solution to the dispute between the two parties. This constructive solution involves a promissory note that the violence will not recur and thus reconciliation is facilitated by this agreement. The state clearly noted that its main responsibility is the preservation of the family unit and that this reconciliatory process will surely serve as a deterrent for future offense.

This choice of action plan is enacted by family cohesion so deeply embedded in both the state and social structure. This position on how to confront domestic violence came up also in my interviews, which is an evidence of a "faithless translation" of the international norm to combat violence against women. The first to point to a problem was one of my interviewees at KAFA, who alerted me that violence against women is not uniformly addressed in Lebanon. Her claim was that even the approach to dealing with violence against women varied: "For example for some organizations they use the negotiation with the husband and this is something internationally proven that you cannot change the abuser if he does not want to change or without treatment. He is not sick but he needs treatment."[28] I was so intrigued by this idea that I chose to ask more activists about their thoughts on "negotiating with an abuser," and to my surprise they mostly seemed to think that it was the first step to take. While one activist tried to strengthen her position by paralleling such an idea to international relations and the need to negotiate with our enemies, the others just pointed to the fact that above all the goal was to preserve the family.[29] They can be viewed as "faithless translators" being co-opted by their own subject position reinforcing the notion that the social rules in Lebanon identify with the norm that a woman's actions and rights are not a reflection of her individuality but more of her familial and communal ties.

Although there are different rules and practices within marriage and divorce across confessions in Lebanon, for the most part, the reality of the Lebanese context broadly speaking is that people resist the idea of breaking up the family and label it as Western. This idea although closely tied to family honor also includes a larger communal reference. It is what some would describe as pride in their national culture and the need to define this way of being as superior to foreign cultural practices. This is not a unique characteristic of Lebanon; however, it is strongly embedded in Lebanese society, and exhibited itself in the interviews and conferences attended. NGOs do not swim against this stream. As one of my interviewees, an academic sociologist and leader of a prominent NGO, proclaimed: "There are certain things we can progress in, open up, and push forward but the basics we hold onto. The idea about preservation of the family is fixed, and this we will never give up on or change. Because in the West they are dying to rebuild the notion of the family because everything has fallen apart."[30] The same interviewee went further on the defensive by pointing out that while in the Arab world and Lebanon violence against women exists it is not as pervasive as it is in the West.

> In the States a woman is being beaten every second of the day. Over here we are in a much better position than on the outside … And where is it that we really see domestic violence in this country; it is mostly seen amongst the rural people that are still following old patriarchal traditions, those that are not educated.[31]

Not only is this a reflection of how women activists juxtapose Western strategies of combating violence against women as in direct conflict with the fundamental values of Lebanon's communal structure it also is evidence of agents' identities and multiple subject positions that allow them to segregate society and the translation of the international norm.

Conclusion

Combating violence against women is addressed by the Lebanese state and its civil society. However, the process the norm undergoes involves confronting filters of translation of a complex society: These translations are steeped in a deeply ingrained patriarchal order where women's rights are translated through a culture of violence and the continued commitment to family honor and cohesion. This framework has given us an explanation to both the lack and variation in implementation of the international norm of combating violence against women in and across Lebanon.

While the international norm on combating violence against women has crossed boundaries into Lebanon's context, its translation has changed the norm in three fundamental ways. First, violence against women gets subordinated to and dissociated from the culture of violence embedded in a warring state. Through civil war and political instability, translating the international norm of combating violence against women becomes more difficult once positioned against an overall culture of violence. Second, violence against women by agents, social, state, and international structures is translated so as not to violate the principle of family preservation. Lastly, both the state and social structures determine how violence against women is translated; namely without circumventing local conceptions of gender and sexuality and their relationship to rights where women's honor, virginity and their relation to the family are the foundations for maintaining these gendered role distinctions. Combating violence against women in this translated context is prevented by the absence of laws that recognize its legitimacy and existence, for example in marital rape, family violence, and in the appropriate and expected sexual behavior of women. It is important for further research to continue to examine the processes international norms undergo as they are translated into a domestic context. The model proposed here can help to explain the varying levels of legitimacy of these norms at the domestic level and the difficulties in achieving a uniform approach to violence against women in different states and regions. UN Women should put an effort into analyzing processes of norm translation at local level. A better understanding of such translation processes improves international governance by exposing the barriers to norm implementation and forging new paths to women's empowerment.

Notes

1 *In-Depth Study On All Forms of Violence Against Women: Report of the Secretary General* (General Assembly document A/61/122/Add.1), 6 July 2006.

2 Jinan Usta, Jo Ann M. Farver and Nora Pashayan, "Domestic Violence: The Lebanese Experience," *Public Health* 121, no. 3 (2007): 208–19.

3 Martha Finnmore and Kathryn Sikkink, "International Norm Dynamics and Political Change," *International Organization* 52, no. 4 (1999): 887–917; Margeret E. Keck and Kathryn Sikkink, *Activists Beyond Borders: Advocacy Networks in International Politics* (Ithaca, NY: Cornell University Press, 1998); Susanne Zwingel, "How Do International Women's Rights Norms Become Effective in Domestic Contexts? An Analysis of the

Convention on the Elimination of All Forms of Discrimination against Women (CEDAW)," PhD diss., Bochum University, 2005.

4 Thomas Risse-Kappen, Steve C. Roppe and Kathryn Sikkink, *The Power of Human Rights: International Norms and Domestic Change* (Cambridge: Cambridge University Press, 1999).

5 Jeffrey Checkel, "Norms, Institutions, and National Identity in Contemporary Europe," *International Studies Quarterly* 43, no. 1 (1999): 83–114; Andrew P. Cortell and James W. Davis Jr., "Understanding the Domestic Impact of International Norms: A Research Agenda," *International Studies Review* 2, no. 1 (2000): 65–87; Zwingel, "How Do International Women's Rights Norms Become Effective in Domestic Contexts?"

6 Susan Bassnett and Harish Trivedi, "Post-Colonial Translation: Theory and Practice," in *Translation Studies*, ed. Susan Bassnett and Harish Trivedi (New York: Routledge, 1999), 2.

7 Ana Lowenhaupt Tsing, "Transitions as Translations," in *Transitions, Environments, Translations*, ed. Cora Kaplan, Joan Wallach Scott and Debra Keates (New York: Routledge, 1997), 255–58.

8 Ibid., 253.

9 Ibid.

10 Lamia Rustum-Shehadeh, "The Legal Status of Married Women in Lebanon," *International Journal of Middle East Studies* 30, no. 4 (1998): 501–19.

11 Suad Joseph, "Teaching Rights and Responsibilities: Paradoxes of Globalization and Children's Citizenship in Lebanon," *Journal of Social History* 38, no. 4 (2005): 1007–26.

12 Ibid.

13 Brian Nordeste, "Negotiating the Human Rights Debate: Re-Imagining Human Rights Beyond the Geopolitical Divide," *Paterson Review: A Graduate Journal of International Affairs* 6 (2005), http://www.diplomatonline.com/pdf_files/npsia/bruno-2.pdf; Lila Abu-Lughod, "Do Muslim Women really Need Saving? Anthropological Reflections on Cultural Relativism and its Others," *American Anthropologist* 104, no. 3 (2002): 783–90.

14 On adultery (487, 488–89), rape, abuse, and impudence (503–6, 513–15, 627–753, 522), and crimes of honor (562).

15 Ministry of Social Affairs (MOSA) and Lebanese Council to Resist Violence Against Women (LCRVAW), *Guide Book for Training Sessions Involving Prevention Against Domestic Violence* (Beirut, 2005).

16 Interview 19, 2008.

17 Interview 8, 2008.

18 Interview 6, 2008.

19 Lama Abu-Odeh, "Crimes of Honor and the Construction of Gender in Arab Societies," in *Feminism and Islam: Legal and Literary Perspectives*, ed. Mai Yamani (New York: New York University Press, 1997): 141–94.

20 Ibid., 142.

21 As of publication Article 562 of the penal code was removed by the parliament after an intense debate with arguments presented by parliamentarians for both removing and leaving the law intact.

22 Interview 2, 2008.

23 Ibid.

24 Interview 5, 2008.

25 Ministry of Social Affairs, Department of Family Affairs, *Request for Information on the Subject of Elimination of Violence against Women*, 19 September 2005, http://www2.ohchr.org/english/issues/women/rapporteur/Lebanon.pdf.
26 Jutta Joachim, "Framing Issues and Seizing Opportunities: The UN, NGOs, and Women's Rights," *International Studies Quarterly* 47, no. 2 (2003): 257.
27 Ministry of Social Affairs, Department of Family Affairs, *Request for Information on the Subject of Elimination of Violence against Women*.
28 Interview 5, 2008.
29 Interviews 8, 12 and 15, 2008.
30 Interview 15, 2008.
31 Ibid.

Part III

Security governance

9 A feminist analysis of UN Security Council resolutions on women, peace, and security

Claudia von Braunmühl

- **UN resolutions on women, peace, and security**
- **Shift of focus?**
- **Systemic limitations to gender mainstreaming**
- **Tensions and perspectives**

Contrary to what we thought we had reason to hope for, the fall of the Berlin Wall did not bring about a peace dividend and the post-Cold War world did not see a demilitarization of politics. Worldwide military structures strive for new purposes and designs and seek to convince a remarkably skeptical public of their value and worth. War and warfare have not been successfully delegitimized, but are (re)cast as part of the normalcy of a nation meeting its noblest obligations. Correspondingly, the concepts of security and threat have seen great changes. Ever since its appearance in the 1990s in development as well as in military discourse, human security serves as a new reference point for war and peace efforts. At the same time as casting development in the highly ambivalent frame of security, human security offers a gateway for social and gender differentiation.[1] This in turn allows the human rights discourse and feminist perspectives to make demands on the security sector. Thus, on national and international levels the security sector sees itself confronted with the gender mainstreaming mandates articulated in the Beijing Platform for Action which all governments had signed and the international women's movements had vigorously lobbied for.

The debate on and the reality of "failed states" and "new wars"[2] brought to the fore two features of the globalized world which both carry a gender dimension: (a) the increased number of states incapable of providing the public good security; and (b) the failure of development to create sufficient jobs and, as a consequence, rising numbers of un- and underemployment. Ever increasing masses of young men, devoid of regular sources of income and deprived of civil forms of functioning as family providers, seek to manifest their masculinity in forms of more or less organized violence. Both may well and all too often do conjure to violent-prone forms of conflict settlement. Where the monopoly of the legitimate use of physical violence of the state is eroded and legal forms of employment or sufficient assets to secure an income are hard to come by, men may easily be drawn into illegal, potentially violent ways of making a living.

There is a growing body of literature on gender and violent conflict and the impact of gender patterns in the various conflict phases. On a conceptual level by and large a feminist concept of gender has gained ground which extends beyond the gender-differentiation of social reality into a normative critique of gender asymmetries and the quest for emancipatory transformation. On an empirical level it is now generally acknowledged that men and women experience violent conflict and war in significantly different ways, and concepts of masculinity and femininity and of a "proper" and "fitting" gender order play a major role in all phases of violent conflict.

In line with United Nations (UN) parlance, conflict phases usually are defined in terms of conflict escalation, open conflict and post-conflict peacebuilding and reconstruction.[3] When the UN Security Council passed its resolutions with regard to gender it did so primarily relating to the post-conflict phase and concerned with "the importance of involving women in all peace-keeping and peacebuilding measures" as noted in Security Council Resolution 1325. By and large the focus on peace processes and the various types of multidimensional peace operations is maintained in the by now five UN Security Council resolutions usually discussed summarily under the heading of "Women, Peace, and Security." These are Security Council Resolutions 1325 (31 October 2000), 1820 (19 June 2008), 1888 (30 September 2009), 1889 (5 October 2009), and 1960 (16 December 2010).[4] After discussing their content and main thrust the question will be raised, if, along with an increased emphasis on sexual violence, a shift to a discourse of victimhood and a loss of agency ascription is discernible. This then obliges us to reconsider gender mainstreaming with a view to its meaning in different settings and its status as an encompassing meta-strategy.

UN resolutions on women, peace, and security

The story of the making of Security Council Resolution 1325 has been frequently recounted, if differently interpreted. There is consensus that it took an as yet unique effort of sustained networking and lobbying between women's and human rights organizations, individual office holders of

UN member states, agencies and networks of women advocates within the UN-system, and the emergence of thematic resolutions to build a case and offer a wording which eventually resulted in the unanimous adoption of the resolution.[5] Women activists and feminists express pride and satisfaction for having successfully moved "'women and armed conflict' on the main agenda"[6] of international security.

Resolution 1325 addresses four areas:

- equal representation and participation of women in all stages of peace processes and in all UN field-based operations, aspiring to a 50/50 gender balance;
- mainstreaming a gender perspective in all activities of peace processes and reconstruction;
- recognition of the particular needs of women and girls and protection from gender-based violence; and
- an end to impunity regarding sexual and other violence against women and girls and the exclusion of such crimes from post-war amnesties.

The actors addressed range from all operating units within the UN system to member states and local conflicting parties.

Resolution 1325 does mention sexual violence as well as the need to protect women and girls and demands an end to impunity for such acts. Its main thrust, however, lies with representation and participation. The resolution carefully avoids essentialism and remains strictly on the level of the practicalities of gendered communication. It stresses the importance for a gender balance of experiences and perspectives to inform and guide the peace process. The resolution has been translated into over 80 languages and has become a key reference point for women's organizations around the world. However, the pace of implementation is felt to be lamentably slow. Even though guidelines, handbooks, check lists, training manuals and the like have been elaborated, gender components are built into the training of peace forces, gender advisors and focal points have been put into place, the actual representation of women in UN-peace formations remains low.[7]

One important reason why male career models and systematic barriers for women in peace and security are allowed to persist is seen in the absence of accountability mechanisms in the resolution. National action plans are hoped to provide a remedy. Nine years after the adoption of Resolution 1325, Resolution 1889 expresses "deep concern about the under-representation of women in all stages of peace processes." Resolution 1889 requests the General Secretary to submit indicators within six months which could serve as a common basis for monitoring the implementation of Resolution 1325, both within the UN system and in member states. By April 2010 a set of altogether 26 "Specific, Measurable, Achievable, Relevant and Time-bound" (SMART) indicators had been elaborated.

Following reports of the Secretary-General on 1325 with a focus on violence against women and the 2006 in-depth report[8] and resolutions adopted by the General Assembly,[9] activities within the UN system were stepped up. The "UN Action against sexual violence in conflict" at the beginning of 2007 united 13 UN agencies under the call "Stop rape now." In February 2008 the UN Secretary-General's campaign *UNiTE to End Violence against Women*, 2008–15 was launched. In June 2008 Resolution 1820 was passed, like the previous one by unanimous vote. While it refers to Resolution 1325, its primary concern is with sexual violence. The resolution "stresses that sexual violence, when used or commissioned as a tactic of war in order to deliberately target civilians or as a part of widespread or systematic attack against civilian populations, can significantly exacerbate situations of armed conflict and may impede the restoration of international peace and security." It reiterates the need to stop impunity, exclude sexual violence from amnesty provisions, and demands a "policy of zero tolerance of sexual exploitation and abuse in UN peacekeeping operations."

The UN Development Fund for Women (UNIFEM) prioritizes the stipulations of Resolution 1820 in the following order:

- sexual violence is recognized as a tactic of war and therefore constitutes an immediate issue of international peace and security;
- as such it requires a security response through sanctions and field staff sensitization;
- all parties to armed conflict are to adopt concrete action regarding protection and prevention to end sexual violence; and
- the importance of women's participation in all processes and measures ending sexual violence in conflict as well as in peace talks is reconfirmed.[10]

The resolution does mention the need to have women participate in peacekeeping structures and processes of the UN, member states, and on the level of the localities of intervention, but its main concern clearly lies with protection. In fact, the European Union (EU) considered protection of women, prevention and response to sexual violence "the weakest pillar of the implementation of Resolution 1325."[11]

A little over a year later, following the Secretary-General's reports on Resolution 1820 and on Resolution 1325, both with a focus on sexual violence, Resolution 1888 was passed and opened on a highly troubled note: Security Council members remained "deeply concerned over the lack of progress on the issue of sexual violence in situations of armed conflict in particular against women and children, notably against girls." The resolution reaffirms the urgent need for action in the area of protection and prevention as "inaction can send a message that the incidence of sexual violence in conflicts is tolerated." It also demands the appointment of "a Special Representative to provide coherent and strategic leadership" through cooperation with the inter-agency "UN Action against sexual violence in conflict." In February 2010 Margot Wallström, former member of the European Commission from Sweden, took up the position of Special Representative on Sexual Violence in Conflict. Passed in December 2010, Resolution 1960 re-endorses the concern "over the slow process in the issue of sexual violence" and requests the Secretary General to report annually. The request relates exclusively to Resolutions 1820 and 1888 and does not include Resolution 1325.

Shift of focus?

In 2002, commissioned by UNIFEM, Ellen Sirleaf, President of Liberia and recent winner of the Nobel Peace Prize, together with Elisabeth Rehn, former Minister of Defense in Finland, evaluated the implementation of Resolution 1325. They criticized the identification of gender with women and girls and the absence of an analysis of gender orders and their dynamics.[12] The observation points to a contradiction which has marked Resolution 1325 from the start and has been somewhat reinforced in the subsequent resolutions. The majority of the wide and manifold efforts pursued under the thematic umbrella of "Women, Peace, and Security" focuses predominantly on the situation of women with little attention to the gender dimensions of organizational structures and policy orientations. Despite available knowledge inspired by feminist theory regarding the gendered terrain of violent conflict, few of these insights inform the Security Council and its resolutions. On the contrary, particularly under the impact of systematic mass rape in Darfur and Congo, a shift towards addressing the victimhood of women and stressing their special need of protection is noticeable. Sexual violence against women moves center stage. So much so that nine years after the adoption of Resolution 1325 and one year after Resolution 1820, Resolution 1889 finds it necessary to "stress(ing) the need to focus not only on the protection of women but also on their empowerment in peacebuilding." Clearly, there is a growing concern that the attention to sexual violence might contribute to a recasting of women in the mold of vulnerability and victims requiring protection and, by the same token, sidelining the claim of empowerment and participation that is crucial in Resolution 1325.

Governmental as well as civil society actors alike consider the five resolutions as forming part of one consistent policy thrust. Yet, as multilateral and bilateral actors engage in implementation processes, we observe a shift of focus. Norway, for example, in 2010 established a Gender, Peace, and Security Unit of two years' duration in the Ministry of Foreign Affairs. The unit, a pre-project in the context of a defense and security sector reform program, welcomes "the increased focus on Resolution 1820 on sexual violence" which is finally and rightly considered a weapon of war, just like other conventional weapons and methods in warfare. This in turn allows "to promote a strong security perspective," a perspective to be defended against "a broad perspective on violence against women, based on the argument that sexual violence is not limited to being a problem only in war." In other words, theories basing their analysis on a continuum of violence might very well be valid, but are felt to be of little use in the specific context of violent conflict and post-conflict activities.[13] If such theories and approaches are taken as a point of reference, "we will end up watering down the available measures and the responsibility of the commanders, the politicians, and the generals, and the whole possibility of holding them accountable to international law disappears."[14] Prevention of sexual violence reduces the subsequent costs of reconstruction and "recruiting women improves the quality and effectiveness of military operations as well as the security of the forces."[15] Thus, focusing on violence against women in conflict is felt to provide a promising strategy successfully to move women from the soft policy area of women's issues to the hard issues of international security. UNIFEM endorses this view: "Positioning sexual violence as a security issue broadens the constituency making it easier to engage security actors for meaningful impact."[16] In addition, such a strategic positioning might draw the support of those states who are unwilling to underwrite the entire human rights- and equality-based gender mainstreaming agenda, but are ready to condemn sexual violence.[17]

Women's organizations and feminist activists are unhappy about the construction of women as a special group in need of protection as it re-victimizes them and jumbles them in the quasi-homogenous category "vulnerable." Peacebuilding actors are likely to lose sight of women's agency[18] and to miss out on interaction with local women's groups and their quiet, yet vital contribution to conflict transformation. The corresponding gender discourse tends to re-animate the old social construct of the male protector. Placing women in a position of vulnerable victims privileges established military and security actors and sidelines participation and empowerment.[19] The logic of patriarchal protection and women's subordination is reinstalled.[20]

By the same token the binary construction of victim and perpetrator all too easily hides the structures of subordination operating within a patriarchal system. The focus lies on injury caused to the victim and "aberrant behavior"[21] on the part of the perpetrator(s) to be attributed to war deprivations and to be prevented, ranged in or disciplined by zero-tolerance strategies and judicial action. The "patriarchal dividend"[22] remains firmly obscured, the systematic advantage that is, which men as a group enjoy. As a consequence, there is no room for empowerment strategies. The projection of women as potential victims suggests protection and submission to the terms coming with it as primary, if not sole solution. The results come across as unconvincing either way. Neither are women effectively protected, as the recent mass rapes in Congo near a UN base in the year 2010 once again demonstrated,[23] nor did the major signatories to the resolutions object when the 2009 Afghan National Stability and Reconciliation Law offered amnesty options that included sexual violence, rape, and abuse.[24]

In short, we see serious discursive collateral damage: the gendered structure of social contexts is sidelined and the construction of masculinities and femininities obscured. The gender dynamics of conflict with militarized masculinity and politicized femininity have little chance to inform analysis and strategy.[25] The "special relationship" between men and weapons and the frequently observed tendency of post-war societies to seek reassurance of "normalcy" in decidedly conservative gender orders are obscured.[26] The end of impunity for gender-based violence repeatedly called for remains on the individual level; there is not even a hint at a societal analysis of the underlying gendered power imbalances.

If we take the five resolutions as one ensemble seeking to gender mainstream the theme of women, peace, and security into international policies, is the concern justified that over a decade a shift from empowerment to protection has occurred? Are concepts of agency and transformative strategies collateral damage of a focus on sexualized violence, without the latter in fact (Darfur, Congo) being combated effectively? The very wording of the reminding and reinforcing resolutions adopted in 2009 could well be read to that effect. They stress both gender mainstreaming and empowerment, with a frequency and urgency absent in the "mother" resolutions. There are many indications that this reflects the experiences and apprehensions of women's organizations and feminist activists, including the fear that civil society actors, i.e., women's organizations themselves, might be somewhat marginalized.[27]

However, there is also a more optimistic view. It attributes high value and relevance to the fact that "women's direct participation and the inclusion of their interests"[28] have been acknowledged as a valid issue on the security agenda. This, in conjunction with the participation stipulations of the resolutions, is considered to open "space where women can assume leadership and political power." Therefore, "despite the usual lip service from official bodies in response to these measures, activists have been able to widen the political space for women's peacebuilding and have developed new paradigms."[29] According to this perspective gender mainstreaming has achieved regulatory effects in that it forms an integral part of security policy debates and has successfully been integrated into the operation directives of all relevant agencies. There are gender statistics, gender training and sensitization. There are efforts to recruit more women into the ranks of peace and security institutions and organizations. And the administrative apparatus of UN security governance is equipped with gender units, gender advisors, focal points and the like. Following this view, for all practical purposes, gender has become a "leitmotif" of UN peace operations.[30]

Systemic limitations to gender mainstreaming

Ever since the adoption of the Beijing Platform for Action in 1995 gender mainstreaming has been read as expressing the claim that women's issues be transformed into societal issues. In interaction with the authorities that has meant that gender mainstreaming has become the primary vehicle and valid instrument to struggle for gender-just discursive, social and political practices. Since then and indeed since long before gender mainstreaming reached the area of peace and security, feminist debates have expressed much disillusionment with the strategy and seriously questioned its value. In its field of origin, development policy, it soon turned out to be an "elusive agenda."[31] In merging an integrative approach with a transformative claim the dominant policy frame continues to write the script with at best rhetorical and statistically marginal concessions to the ill-fitting gender quest. Many women's organizations and feminist activists seek a profound revisiting of gender mainstreaming. Transformative strategies require a radical analysis of global power structures and have to include the realization that gender power relations are a predisposing cause to the use of violence.[32]

In the area of peace and security there are two important aspects of immediate and direct pertinence to gender mainstreaming which are discussed surprisingly little. These regard (a) the nature and quality of participation; and (b) the nature into which the stream gender is to be integrated or that it is supposed to transform, that is, the liberal peace agenda.

Participation

With gender mainstreaming, a social movement addresses the mechanisms of state agencies and, in the area of peace and security, does so in a realm steeped in patriarchal, hierarchical thinking in practice and theory alike.[33] The very sexual contract that Carole Pateman[34] analyzes as lying at the base of state formation shapes, in structure and prevailing attitudes, the actors called upon to provide protection and to respect equality. Androcentric features and a male work culture shaping the inner life of institutions and organizations certainly work, by all accounts are even accentuated, in the peace and security sector. The dynamics of decision making within institutions and their operating mode is deeply entrenched with the patriarchal gender order. Gender hierarchies and male-bonding mold prevalent norms, criteria for bestowing legitimacy, procedural requirements, career patterns, achievement criteria, expectations and informal cultural mores. In the field of peace and security even more so than in others it is virtually impossible to expose inner-institutional blockages and to take recourse to external movement constituencies or to the general public. This constitutes a major impediment to meaningful participation at the same time that it tends to individualize and silence relevant experiences made within institutions.

The debate revolving around the five women, peace, and security UN resolutions appears to take participation per se as an, as it were, self-guaranteed stepping stone to personal empowerment and transformative influence. Yet, it is by no means evident that the encounter between women and the security apparatus generates empowering interaction and negotiations and that a gender perspective stands a chance. In fact more often than not participation comes with a high price, undermining original political positions, eroding personalities or simply gatekeeping gender advocates in "their" niche.[35] The functional arguments for participation—e.g., women get easier access to local women and by implication to local gender realities—may well be true. But there is also truth in the repeated, if anecdotal, evidence we receive from gender advisors in military and police settings on the difficulties they face in getting access to the higher echelons they are to report to. Finally, representation of women by quantitative participation is not necessarily identical with power of voice. While the statistical record simulates participation, in fact women in peace consultations may be expected

(Somalia) or ordered (Afghanistan)[36] to remain silent. These observations are not meant to denigrate the strife for increased representation and participation of women in all stages of peace processes, but they do point to the need to approach the actors on all sides through the critical lens of an in-depth gender analysis.

Liberal peace

Feminist development discourse has long moved to a critical in-depth analysis of corporate-driven globalization with its social polarization on international and national levels, eroding the fabric of societies and destabilizing governments and states. While in continuous communication and interaction with feminists in academe, much of the critical theory-building derived from the experience of working inside and with the women's machineries created during and after the UN decade of women in aid agencies as well as in recipient countries. Increasingly programs and policies designed to "fight poverty" target women, e.g., by virtue of the much celebrated micro-credits. In the process the term "empowerment" was reframed by aid agencies and became synonymous with economic empowerment. Introduced into the politics of the international women's movements in 1985 by Development Alternatives with Women for a New Era (DAWN), a network of feminist activists and researchers, as a marker for an emancipatory process, empowerment has since come to denote the self-management of the marginalized.[37]

Therefore, to the women's movements in developing countries gender mainstreaming quickly has become part and parcel of the neoliberal agenda of macroeconomics and macropolitics of structural adjustment. This is in stark contrast to the world of the Organisation for Economic Co-operation and Development (OECD), certainly to Europe, where gender mainstreaming encountered and was absorbed into an equality discourse and the fairly developed institutional mechanisms of the member countries of the EU. This discourse to this day does not necessarily take issue with the economic and political mainstream, but rather focuses on making inroads into gender imbalances. However, more recently feminist critics of neo-classical political economy[38] have raised questions concerning the kind of peace multi-dimensional peacemaking is pursuing and the nature of its making. In this perspective the concepts of peace and reconstruction are analyzed as framed by the dominant globalization agenda, prescribing certain strategies and excluding others. As much as liberal peace and post-war reconstruction are vividly debated in the peace research and in the development community, by all appearances the issue has not yet figured prominently in the Women, Peace, and Security discourse based on the five Security Council resolutions.

There are in fact critical issues to be dealt with. The (neo)liberal grammar underlying the liberal peace agenda spells out the guiding concepts of human rights and citizenship, democracy and free market, civil society and government, nation and state, and their place in the international order. This reading of the human rights concept prioritizes individual rights and political citizenship with elections as a centerpiece of political reconstruction, downplaying social and economic rights in the process. Usually, the transition into a post-conflict situation is not accompanied by reforms designed to undo or only curtail the—formal and informal, legal and illegal—socio-economic power structures, alliances and networks that the war has generated and fuelled. As a consequence, warlords and war profiteers re-enter the scene in the shape of successful business men and politicians ready to profit from open markets with their well-established global connections.

The liberal peacebuilding paradigm with its projection of a threefold transformation to peace, democracy and market economy as a self-sustaining process is reminiscent of the one-size-fits-all approach of structural adjustment. Under the circumstances locally resonant forms, strategies, and visions of peacebuilding which women often engage in may well find it difficult to be acknowledged.[39]

Tensions and perspectives

We do not find much evidence that the debate on Resolution 1325 and the following resolutions goes beyond the discursive space delineated by the paradigm of liberal peace. The terms framing reconstruction and recovery are rarely the subject of discussion. The fact that the liberal frame remains beyond debate or even challenge may well be due to a combination of genuine common ground and a heavy dose of seeking connectivity to the dominant model of development and to those who wield power in it. Quite possibly the silence is driven by the genuine desire to secure empowerment and transformation through participation and protection.[40] This raises the question: Is the Women, Peace, and Security agenda condemned to navigate within the dominant neoliberal frame in a similar fashion as was suggested by Anne Sisson Runyan[41] more than 10 years ago referring to the Beijing agenda? Is the gender mainstreaming debate surrounding the five resolutions inadvertently, but necessarily, part and parcel of a hegemonic claim patterned on far-reaching but narrow-minded universalist aspirations? Is the shift to victimhood and protection not only responding to the increase of sexualized violence in conflict regions, but also conducive to a tacit compromise with the conservative images of society and gender order underlying the neoliberal vision of the world? Is the complicity with the dominant agenda that much of gender mainstreaming on European and national levels has evolved into re-enacted on international levels? Can gender mainstreaming in peace and security find entry points into institutional practices other than the windows of opportunity offered, or rather permitted, by a human resources management that seeks to optimize gender-specific differences for the purposes of increasing organizational effectiveness and efficiency? The debate has only just begun and is far from having settled into a unifying consensus.

One line of argument demands to reset the debate in the continuum of violence ranging from societal structures of discrimination and subordination to violence in private and public spheres to war. Even though the types of violence in each context vary greatly and in terms of damage inflicted may defy comparison, on a structural level one predisposes the other. Sexual violence against women as an instrument of war would not be possible without women being objectified, denigrated and suppressed in civil life.[42] On a normative and theoretical level this is entirely consistent with the gender relevance of the human rights frame reaffirmed by the World Conference on Human Rights in 1993. At that occasion it was stated in clear terms that the entire human rights agenda safeguarding the dignity and the physical integrity without limitations applies to women and the specific forms of violation of human rights they may encounter. On a practical level, the exigencies and conditioning factors of focused action tend to turn out much more limited. This certainly is the experience in development policy. In the 1990s Gender in Development theory began to subscribe to the differentiation between practical and strategic gender needs.[43] In the struggle for development interventions to actually meet the practical needs of women time and again strategic issues are sidelined, if not warded off altogether for allegedly dangerously overloading the agenda.

It is precisely this continuum and the politically unmanageable width of the axis which is reflected in past and present debates and bifurcations. Ever since the 1970s, peace movements have been influenced by Johan Galtung's juxtaposition of structural and direct violence and the corresponding notions of positive and negative peace.[44] We see the resurgence of that same juxtaposition in the human security debate in the form of a wider concept of human security encompassing whole systems of well-being as first posited by the UN Development Program (UNDP) in its 1994 Human Development Report, and a narrower concept confined to physical integrity.[45] While on a theoretical level the concepts of structural violence, positive peace and social well-being are of great analytical and normative value, when it comes to strategizing and defining action they may be far too interventionist.

At this juncture, critical peace researchers remind us of the fact that the local and its political, economic, social, customary, cultural and spiritual dynamics are all too often ignored. Whether speaking from a post-colonial perspective[46] or coming from a social justice tradition[47] they agree on the need for more space for local agency and strategies more resonant with the social and political visions shared on local levels. Drawing on Michael Walzer's[48] distinction between thick and thin moral arguments in defense of a just war, Richmond suggests "thin cosmological norms and thick local expectations from where a social contract must emanate."[49] Peacebuilding must allow for the space to accommodate emergent local-liberal forms of hybridity including local concepts of political legitimacy and forms of organizing political power different from the Western model. In order to do so, it must proceed with great analytical sharpness and practical flexibility.

Unfortunately there has been remarkably little dialogue between critical peace studies and their analysis of the pitfalls of liberal peace with social movements, in particular women's organizations, and their claims for inclusive citizenship.[50] The sweeping dismissal of a universalist frame of reference as part of a hegemonic design ignores that in many ways the conflict between universal reach and local significance, between the historical roots and present validity of human rights (*Genesis und Geltung*) has long been resolved. The women's movements of the world for once have insisted on the human rights frame as relevant for their struggles against subordination of whichever shape and reasoning. In given power settings and embedded by actors of peacebuilding who undoubtedly subscribe to the liberal peace agenda it is indeed extremely difficult to find out what "local-liberal forms of hybridity" could bring to the agenda of rebalancing subordinating gender relationships.

On a more general note, much of the disappointment with gender mainstreaming is brought about by high expectations, if not a strategic overload, which it may be time to review. The Beijing Platform for Action lays down essential objectives of international women's policy and, as a matter of methodological procedure, requires from all those in authority systematic self-information on gender effects and gender-differentials of their actions. The wording does not suggest that gender mainstreaming as such is expected to encompass all the objectives of the platform. As a strategic and tactical move it is certainly legitimate to charge gender mainstreaming to the fullest. But there does remain an obvious tension between integrative approach and transformative claim, possibly stronger than anywhere else in the male, hierarchic world of established security actors.

Perhaps the time has come to consider a strategic unbundling. Perhaps we ought to take a step back and reconsider what gender mainstreaming can achieve in general and in peace and security in particular, and where, without abandoning or discrediting gender mainstreaming, complementary or independent strategies are called for. The role that UN Women should play in this re-consideration is to insist on a concept of human rights that under all circumstances takes issue with the violations of women's physical integrity and well-being. Such violations and infringements are embedded in deeply entrenched gendered power imbalances and need to be confronted unwaveringly.

Notes

1 Mark Duffield, *Development, Security and Unending War* (Cambridge: Polity Press, 2007); Claudia von Braunmühl, "Human Security—ein Diener vieler Herren," in *Kontroversen über den Zustand der Welt*, ed. Marcel van der Linden and Christoph Lieber (Hamburg: VSA, 2007), 181–98.

2 Mary Kaldor, *New and Old Wars: Organized Violence in a Global Era* (Cambridge: Polity Press, 1999); Sven Chojnacki, "The Continuum of Violence," in *Sites of Violence. Gender and Conflict Zones*, ed. Wenona Giles and Jennifer Hyndman (Berkeley, CA: University of California Press, 2004), 24–44; Peter Lock, *War Economies and the Shadow of Globalisation* (Berlin, Germany: Berghof Research Center for Constructive Conflict Management, 2005), http://www.berghof-handbook.net/uploads/download/dialogue3_lock.pdf.

3 For example, *In-depth Study on All Forms of Violence Against Women: Report of the Secretary-General* (General Assembly document A/61/122/Add. 1), 6 July 2006; Rashida Manjoo, *Report of the Special Rapporteur on Violence Against Women, its Causes and Consequences* (General Assembly Document A/HRC/14/22), 23 April 2010, http://www2.ohchr.org/english/bodies/hrcouncil/docs/14session/A.HRC.14.22.pdf; Claudia von Braunmühl, "Geschlechterdimensionen gewalttätig ausgetragener Konflikte in der Internationalen Politik" in *Gender … Politik … online, Gender in den Sozialwissenschaften* (January 2008), http://web.fu-berlin.de/gpo/claudia_braunmuehl.htm.

4 The five resolutions can be accessed at http://www.peacewomen.org/themes_theme.php?id=65&subtheme=true.

5 Laura J. Shepherd, "Power and Authority in the Production of United Nations Security Council Resolution 1325," *International Studies Quarterly* 52 (2008): 383–404; Torunn L. Tryggestad, *Trick or Treat? The UN and Implementation of Security Council 1325 on Women, Peace and Security* (2009), http://www.thefreelibrary.com/Trick+or+treat%3F+The+UN+and+implementation+of+security+Council…-a0215069791.

6 Cynthia Cockburn, *From Where We Stand. War, Women's Activism and Feminist Analysis* (London: Zed Books, 2007), 140.

7 For detailed data see Veronica Haglund, "Wanted: More Women to Keep the Peace," *UNA USA World Bulletin*, 20 January 2010, http://www.unausa.org/worldbulletin/012010/haglund; Peace Research Institute Oslo (PRIO), *Gender, Peace and Security Update*, April 2010, http://www.peacewomen.org/assets/file/Resources/Academic/priogpsapril_issue_2010.pdf and May-June Issues 2010, http://www.prio.no/sptrans/-553465591/PRIO%20Gender,%20Peace%20and%20Security%20Update%20May-June%20Issue.pdf.

8 *Statement by the President of the Security Council* (Security Council document s/PRST/2004/40), 28 October 2004, http://www.un.org/womenwatch/ods/S-PRST-2004-40-E.pdf; *In-depth Study on All Forms of Violence Against Women: Report of the Secretary-General.*

9 *Intensification of Efforts to Eliminate All Forms of Violence Against Women* (General Assembly document 61/143), 30 January 2007, http://www.uneca.org/daweca/conventions_and_resolutions/Res.%20intensification%20of%20efforts.pdf; *Intensification of Efforts to Eliminate All Forms of Violence Against Women: Report of the Secretary General* (General Assembly document A/64/151), 17 July 2009, http://daccess-dds-ny.un.org/doc/UNDOC/GEN/N09/407/73/PDF/N0940773.pdf?OpenElement, 19 December 2006, http://reliefweb.int/sites/reliefweb.int/files/reliefweb_pdf/node-328764.pdf.

10 UNIFEM, *Security Council Resolution 1820. Women, Peace and Security* (2010) http://www.unifem.org/gender_issues/women_war_peace/unscr_1820.php.

11 *Comprehensive Approach to the EU Implementation of the United Nations Security Council Resolutions 1325 and 1820 on Women, Peace and Security* (Council of the European Union document 15671/1/08 REV 1), 1 December 2008, 9, http://www.consilium.europa.eu/ueDocs/cms_Data/docs/hr/news187.pdf.

12 Elisabeth Rehn and Ellen Johnson Sirleaf, *War and Peace – The Independent Expert's Assessment on the Impact of Armed Conflict on Women and Women's Role in Peace-Building* (New York: UNIFEM, 2002).

13 For a continuum of violence approach see, e.g., Caroline O.N. Moser and Fiona C. Clark, ed., *Victims, Perpetrators or Actors: Gender, Armed Conflict and Political Violence* (London: Zed Books, 2001); Amie Alden, "A Continuum of Violence: A Gendered Analysis of Post Conflict Transformation," *POLIS Journal* 3 (Winter 2010).

14 Peace Research Institute Oslo (PRIO), *Gender, Peace and Security Update*, April and May–June, 2 (2010).

15 Ibid., 3.

16 UNIFEM, *Security Council Resolution 1820. Women, Peace and Security* (2010), 4, http://www.stoprapenow.org/uploads/advocacyresources/1282164625.pdf.

17 Jan Arno Heßbrügge, "Frauen im Krieg, Krieg gegen Frauen. Die Vereinten Nationen und der Kampf gegen sexuelle Gewalt in bewaffneten Konflikten," *Vereinte Nationen* 5 (2009): 212–16. Two aspects are neglected here: 1) Sexualized or gender-based violence tends to be thought of as violence inflicted by men on women. Sexual violence between men in the arena of war, by all appearances perceived as act of feminization, is shrouded in potent taboos and much less documented. 2) Women are not exclusively on the receiving end of violence. They may in many ways seek access to the actors of violence, approve of their action or even participate in violent acts themselves. See von Braunmühl, "Geschlechterdimensionen gewalttätig ausgetragener Konflikte in der Internationalen Politik."

18 Sanam Naraghi Anderlini, *Women Building Peace: What They Do, Why It Matters* (Boulder, CO: Lynne Rienner, 2007).
19 Donna Pankhurst, "The 'Sex War' and Other Wars: Towards a Feminist Approach to Peace-building," in *Development, Women and War: Feminist Perspectives*, ed. Haleh Afshar and Deborah Eade (Oxford: Oxfam, 2004), 8–42; Heßbrügge, "Frauen im Krieg, Krieg gegen Frauen," 212–16.
20 Iris Marion Young, *Global Challenges. War, Self Determination and Responsibility for Justice* (Cambridge: Polity Press, 2007), 117. An inkling of the unease seems to come through when the Secretary-General's 2009 report on the implementation of resolution 1820 in a footnote painstakingly explains why, after having given it some thought, the term "victim" is after all preferred to that of "survivor." *Report of the Secretary-General Pursuant to Security Council Resolution 1820* (Security Council document S/2009/362), 20 August 2009, 3, http://daccess-dds-ny.un.org/doc/UNDOC/GEN/N09/405/56/PDF/N0940556.pdf?OpenElement.
21 Ra'a Zeid Hussein Zeid, *A Comprehensive Strategy to Eliminate Future Sexual Exploitation and Abuse in United Nations Peacekeeping Operations* (General Assembly document A/59/710), 24 March 2005, 20, http://cdu.unlb.org/Portals/0/Documents/KeyDoc5.pdf.
22 Raewyn Connell, *Gender* (Cambridge: Polity Press, 2002), 142.
23 Jeffrey Gettleman, "Mass Rapes in Congo Reveals U.N. Weakness," *New York Times*, 3 October 2010.
24 Medica Mondiale, "'Wir waren voll Hoffnung auf eine bessere Zukunft … ,' Positionspapier von zur Lage der Frauen in Afghanistan, Juni 2010" (June 2010), http://www.medicamondiale.org/fileadmin/content/07_Infothek/Positionspapiere/Positionspapier_Afghanistan–mediica_mondiale_Juli_2010.pdf.
25 Michael Kimmel, "Globalization and its Mal(e)contents. The Gendered Moral and Political Economy of Terrorism," *International Sociology* 18, no. 3 (2003): 603–20; Henri Myrttinen, "Disarming Masculinities," *Disarmament Forum: Women, Men, Peace and Security*, no. 4 (2003): 37–46.
26 Von Braunmühl, "Geschlechterdimensionen gewalttätig ausgetragener Konflikte in der Internationalen Politik."
27 In a consultation in March 2009 Charlotte Bunch, head of the Center for Women's Global Leadership and leading activist in the struggle for Resolution 1325, regretted that civil society had been virtually excluded from the drafting of the Framework of Action of the Secretary General's campaign "Unite to end violence against women." The fact that the statement by the Commission on the Status of Women concerning Resolution 1325 passed on International Women's Day 2010 was decided upon without any prior civil society consultation is also taken as pointing in that direction; see UNIFEM and UN Inter-Agency Working Group of the Secretary-General's Campaign, *Consultation with Civil Society on the Secretary-General's Campaign UNiTE to End Violence Against Women, Summary Report and Recommendations*, 3 March 2009, 3, http://www.stoprapenow.org/advocacy-resources/index/?t=10&p=3.
28 Tryggestad, "*Trick or Treat?*" 7.
29 Shelley Anderson, "Vital Peace Constituencies," 18 February 2010, http://www.opendemocracy.net/comment/reply/50366.
30 Ulrike Baumgärtner, "Gleichstellung der Geschlechter in der UN-Verwaltung im Bereich der Friedenskonsolidierung," *Sicherheit und Frieden* 28, no. 1 (2010): 13–20, 19.

31 Rounaq Jahan, *The Elusive Agenda: Mainstreaming Women in Development* (London: Zed Books, 1995); Anne-Marie Goetz, ed., *Getting Institutions Right for Women in Development* (London: Zed Books, 1998).

32 Cynthia Cockburn, "Gender Relations as Causal in Militarization and War. A Feminist Standpoint," *International Feminist Journal of Politics* 12, no. 2 (June 2010): 139–57.

33 Spike V. Peterson, *Gendered States. Feminist (Re)Visions of International Relations Theory* (Boulder, CO: Lynne Rienner Publications, 1992); Ann J. Tickner, *Gender in International Relations: Feminist Perspectives on Achieving Global Security* (New York: Columbia University Press, 1992).

34 Carole Pateman, *The Sexual Contract* (Cambridge: Polity Press, 1988).

35 At an official function on Resolution 1325 in Berlin designed to present career role models to a public of mostly young women one of the few (former) female heads of a peace negotiation mission described in impressive detail how she was marginalized, cut-off from the flow of interaction and information and mobbed into a position of servitude. Her career survived, and this is why she presented her experience as a positive example.

36 In Somalia UNDP funding for peace conferences was made conditional on the participation of women. The warlords brought in wives and relatives to enjoy the luxury of conference hotels and be silent. Similarly in Afghanistan warlords bowed to the participation stipulations of the Bonn Agreement, but, back in Kabul, commanded female representatives to be prudent—and quiet (participant's report).

37 Maxine Molyneux and Shahra Razavi, ed., *Gender Justice, Development, and Rights* (Oxford: Oxford University Press, 2002).

38 Nancy Fraser, *Scales of Justice: Reimagining Political Space in a Globalized World* (New York: Columbia University Press, 2009); Chandra Talpade Mohanty, "Under Western Eyes Revisited: Feminist Solidarity through Anticapitalist Struggles," *Signs, Journal of Women in Culture and Society* 28, no. 4 (2002): 499–535; Diane Elson, "Gender Justice, Human Rights and Neo-liberal Economic Policies," in *Gender Justice, Development and Rights*, ed. Maxine Molyneux and Shahra Razavi (Oxford: Oxford University Press, 2002), 78–114.

39 Sabine Kurtenbach, "Why is Liberal Peace-building so Difficult? Some Lessons from Central America," *GIGA Working Papers*, no. 59 (September 2007), http://www.conflictrecovery.org/bin/Why_is_liberal_peace-building_so_difficult-GIGA_wp59_kurtenbach.pdf; Oliver Richmond, "Liberal Peace Transition: a Rethink is Urgent," 19 November 2009, http://www.opendemocracy.net/oliver-p-richmond/liberal-peace-transitions-rethink-is-urgent.

40 Shepherd, "Power and Authority in the Production of United Nations Security Council Resolution 1325," 383–404.

41 Anne Sisson Runyan, "Women in the Neoliberal Frame," in *Gender Politics in Global Governance*, ed. Mary K. Meyer and Elisabeth Prügl (New York: Rowman & Littlefield Publishers, 1999), 210–20.

42 Moser and Clark, *Victims, Perpetrators or Actors?*; Cockburn, *From Where We Stand*; Cathy McIlwaine, *Multi-scalar Interpretations of the Gendered Continuum of Violence: Reflections from Columbia and Guatemala* (Mimeo, 2010).

43 Caroline O. Moser, *Gender Planning and Development: Theory, Practice, and Training* (New York: Routledge, 1993).

44 Johan Galtung, "Violence, Peace, and Peace Research," *Journal of Peace Research* 3 (1969): 167–91.
45 Von Braunmühl, "Human Security."
46 For example, Oliver Richmond, "Resistance and the Post-liberal Peace," *Millennium, Journal of International Studies* 38, no. 3 (2010): 665–92, http://www.peace-building.de/files/mil365017_665_692.pdf.
47 For example, Kurtenbach, "Why is Liberal Peace-building so Difficult?"
48 Michael Walzer, *Thick and Thin: Moral Argument at Home and Abroad* (Notre Dame: Notre Dame Press, 1994).
49 Richmond, "Resistance and the Post-liberal Peace."
50 Naila Kabeer, ed., *Inclusive Citizenship: Meanings and Expressions* (London: Zed Books, 2005).

10 European countries and the implementation of UN Security Council Resolution 1325

Andrea Schneiker and Jutta Joachim

- **Numerical representation of women**
- **Gender training**
- **Protection of women from gender-based violence**
- **The role and needs of women related to peace agreements**
- **Analysis**
- **Conclusion**

In the fall of 2000, members of the United Nations (UN) Security Council adopted Resolution 1325 (SCR 1325) entitled "Women, Peace and Security."[1] Nongovernmental organizations (NGOs) and states alike considered the document as revolutionary at the time. In light of increasing numbers of incidents of sexual abuse as well as harassment perpetrated by UN peacekeepers, on the one hand, and given the growing violence against women in ongoing civil wars, on the other hand, the resolution recognizes "the urgent need to mainstream a gender perspective into peacekeeping."[2] Since then, the UN, regional organizations (e.g., the European Union (EU), the Organization for Security and Co-operation in Europe (OSCE), or the North Atlantic Treaty Organization (NATO)), and individual governments have started to implement and apply the resolution. While the Security Council invites national governments to establish national action plans (NAPs) in order to implement the provisions of the resolution,[3] non-adoption is not sanctioned. As of September 2011, only 26 member states had adopted an NAP.[4]

SCR 1325 has prompted scholars to take both a closer look at the relation between gender and peacekeeping[5] and NAPs as means of implementing the resolution.[6] Nevertheless, there is a void with respect to comparative studies that systematically examine how the resolution has been translated into national politics by UN member states.[7] In this chapter, we therefore discuss the findings regarding the implementation of SCR 1325 in three different countries: Germany, the United Kingdom (UK), and Sweden. These three cases are of particular interest. First, national action with respect to SCR 1325 is, in part, inconsistent with what these countries do internationally. Although each of the three countries has been a leader in bringing about the resolution and diffusing it to other regional organizations, national implementation in these countries has proceeded rather differently. While the United Kingdom and Sweden have both adopted an NAP and engaged in a whole range of gender mainstreaming activities in both the military and civilian sector, Germany has no such plan as of yet and, consequently, lacks a coherent implementation strategy. Second, considering that each country that takes part in international peacekeeping missions, and has to implement SCR 1325, is responsible for conducting its own gender training,[8] it is not only relevant to study how the resolution is implemented in countries where conflicts take place, but also what governments do to prepare their troops for peacekeeping missions. Given that there is a lack of gender sensitivity on the part of many peacekeepers, gender trainings are crucial, especially when considering the serious problems in conflict and post-conflict settings where trafficking and sexual violence are only at the tip of the iceberg.[9]

SCR 1325 aims at the mainstreaming of a gender perspective in peacekeeping operations. Gender mainstreaming promotes "gender equality through [the] systematic integration of [gender in] all systems and structures, into all policies, processes and procedures, into the organization and its culture, into ways of seeing and doing."[10] To this aim, the resolution contains a number of important provisions. Out of those, we considered four of the most important ones for our analysis:

(1) Numerical representation of women:
The resolution "[*u*]*rges* Member States to ensure increased representation of women at all decision-making levels in national, regional and international institutions and mechanisms for the prevention, management, and resolution of conflict."

(2) Gender training:
SCR 1325 asks member states to carry out training "on the protection, rights and the particular needs of women" and "HIV/AIDS awareness training […] for military and civilian police personnel in preparation for deployment."

(3) Protection of women from gender-based violence:
The resolution "[*c*]*alls on* all parties to armed conflict to take special measures to protect women and girls from gender-based violence, particularly rape and other forms of sexual abuse."

(4) Role and needs of women related to peace agreements:
It calls on all actors involved to adopt a gender perspective "when negotiating and implementing peace agreements," thereby supporting "local women's peace initiatives," ensuring "the protection of and respect for human rights of women and girls, particularly as they relate to the constitution, the electoral system, the police and the judiciary," and their special needs with respect to disarmament, demobilization and reintegration.[11]

In the following sections we analyze and compare how Germany, the United Kingdom, and Sweden have responded thus far to these four provisions. Though cognizant of their engagement regarding SCR 1325 in international organizations (IOs), we examine activities at the national level only. In the case of the United Kingdom and Sweden, we have analyzed the NAPs and related governmental documents. The United Kingdom adopted the first NAP in March of 2006[12] and published a revised version in November of 2010.[13] Similarly, the Swedish government replaced the first NAP released in October 2006[14] with an updated version in 2009.[15] With respect to Germany, we relied on three reports that the government has prepared on the implementation of SCR 1325[16] and related governmental documents. In addition to the NAPs and the reports, we have analyzed other relevant documents of governmental actors. In the following, we will show how SCR 1325 has been implemented so far, offer explanations for the observed differences, and draw conclusions with respect to the domestic factors that influence norm implementation.

Numerical representation of women

When looking at the number of women in the armed forces, hardly any differences seem to exist between the three countries. In the *Bundeswehr*, the German military, women make up a total of 8.6 percent and account for 5.4 percent in the fighting troops;[17] about 5 percent take part in international missions.[18] In the United Kingdom, the total number of women in the Armed Forces amounts to 9.1 percent,[19] compared to 10.6 percent in Sweden.[20] These low figures have been subject to criticism by activists who consider them as too low to reach a critical mass necessary to bring about changes[21] and for gender mainstreaming to have any substantial impact.

Although the governments of all three countries have declared in line with SCR 1325 that they seek to increase the number of women in the armed forces and in other national institutions that are involved in the prevention and resolution of conflict, they have done so in different ways. The German government "especially encourage[s] women to apply for open positions within international organizations and to address them directly in particular cases"[22] to increase the number of women with respect to peacekeeping missions. The first United Kingdom NAP as well as that of Sweden, by comparison, is reflective of a more proactive approach. The UK government plans to "[l]obby through the UN and UK Missions overseas, to encourage Member States to identify suitable female candidates," to "[s]end e-gram with instructions to all diplomatic posts and HMG [Her Majesty's Government] desk officer stakeholders to increase awareness of the importance of increasing the numbers of women involved in conflict resolution /peace-building," and to "[e]stablish a virtual network of key female personnel to disseminate vacancies quickly and widely."[23] Similarly, in Sweden, the armed forces and the police have "implemented special initiatives to increase the proportion of women" able to participate in international peacekeeping missions. The NAP states, on the one hand, that "[t]he relevant authorities will be instructed to review job descriptions, working environment issues and recruitment models and difficulties, as well as to compile relevant statistics and take additional measures," and, on the other hand, that "[m]ore women candidates will be identified and nominated for high positions in international operations."[24] The second NAP even requires that "[i]n civilian peace-keeping and security-building operations, and in the civilian components of multifunctional operations, there is to be equal participation of women and men" and that in military and police peacekeeping and security-building operations "women and men are to participate on equal terms."[25] This contrasts with Germany, where the overall goal is to reach quotas of 50 percent female soldiers in the medical service and only 15 percent in combat troops.[26]

Gender training

With respect to gender training of military personnel taking part in international missions, we also found significant differences across the three cases. In its 2007 report, the German government identifies a program it has developed to assist German soldiers in interactions with the other sex and which focuses, among other factors, on "the compatibility of family life and profession"[27] as satisfying the provisions of SCR 1325. However, it does not indicate how the mentioned program relates to issues of peace and security and, hence, is of relevance for the implementation of SCR 1325.

Its relevance becomes only somewhat more clear in a working paper of the military entitled "Deciding and taking responsibility—conflict situations in international missions." The paper discusses how German soldiers should handle "need, misery and oppression of the civilian population" in conflict areas.[28] In addition to measures that focus on the internal organization of the army and the relations among soldiers, the third report of the German government on the implementation of SCR 1325 that dates from 2010 explains that gender aspects, including the role of women in conflict resolution, are incorporated in the training for personnel that is about to take part in UN missions.[29] Yet, the report does not specify who exactly is to be trained, what the training elements ought to be, who is carrying out the training, how long the training lasts, nor if it is mandatory.

The second NAP of the UK government, by comparison, states as a goal that "[a]ll staff working on conflict issues across FCO [Foreign and Commonwealth Office], MoD [Ministry of Defence] and DfID [Department for International Development] departments are aware of the women peace and security agenda and receive specialist training, when appropriate."[30] Working towards it, the NAP envisions different training modules, identifies individual governmental agencies responsible for their implementation, and lists indicators to measure their successful implementation. Of the different training modules, equality and diversity training for the UK armed forces is considered mandatory.[31] In addition, some female soldiers receive, for example, training in preparation of their deployment as part of so-called Female Engagement Teams (FETs). Stationed in each operational area,[32] the teams are "made up of female soldiers and have been developed to enable the military to better engage with communities in Afghanistan." The training comprises different elements and includes, for example, "advice on Afghan culture and specifics relating to Afghan women."[33]

Similar to the UK NAP, the Swedish plan places great emphasis on training: "Swedish personnel who take part in peace support […] operations will have undergone training (or in special cases demonstrated knowledge in some other way) that includes Resolution 1325."[34] While the government has assigned responsibility to individual agencies to ensure and carry out the training, it also relies on the expertise of civil and military, state and non-state actors. The "Gender Field Advisor" course, for example, "is organized in collaboration between the Swedish Armed Forces, the Swedish Civil Contingencies Agency (MSB), the Swedish Police, the Folke Bernadotte Academy and the Swedish non-governmental organization Woman to Woman."[35] It provides, among other things, "knowledge and background of UNSCRs 1325, 1820,

1888 and 1889; gender perspectives in operations/missions; experiences and lessons learned from previous gender field advisers; methods, policy and best practices; cooperation with other actors internal and/or external (NGOs, IOs); general knowledge about gender in relation to UN, EU, NATO and OSCE documents; policies, directives; gender as an operational tool and code of conduct." In the field, the gender field advisor "is the head of mission's/commander's greatest resource in the implementation of gender awareness into any operation."[36]

With respect to training, the Swedish government benefits from the resources of quite a unique institution, the Folke Bernadotte Academy, which is a "government agency dedicated to improving the quality and effectiveness of international conflict and crisis management, with a particular focus on peace operations."[37] The academy seeks to encourage and facilitate cooperation not only between different national institutions, but also between the latter and their international partners.[38] Training is an important component of the academy's work and it offers a variety of gender training courses. In addition to the one for gender field advisors, it also conducts a course related to "Leadership and Gender in Peace Operations," which is premised on the idea that "[g]ender affects both the internal and external environment of peace operations" and that, "[t]herefore, gender awareness is key to effective leadership."[39] Although special courses on gender exist, the academy "aims to integrate a gender perspective into all of its courses and exercises."[40] Furthermore, its training courses are not limited to Swedish personnel, but are also open for international participants.[41] In light of the ongoing activities and compared to Germany and the United Kingdom, Sweden appears to possess the most sophisticated infrastructure and training programs when it comes to gender training in the context of SCR 1325.

Protection of women from gender-based violence

With respect to the measures that shall ensure the protection of women and girls from gender-based violence, the military as well as the civilian branch of the German government appear to have implemented SCR 1325 independently of each other since they advocate different strategies. Relevant publications of governmental development aid agencies related to gender-based violence in the context of conflicts, in general, and security sector reform, more specifically, are rather telling in this respect.[42] In one of them, entitled "Gender and Conflict," a gender perspective is considered important for conflict resolution. Moreover, the document discusses not only the role of women with respect to violence and conflict resolution, but also that of men and masculinities.[43] In comparison, the military branch of government mentions gender-based violence only in connection with training of German military personnel prior to peacekeeping missions and their interactions with the local population.[44] In addition, whereas the perspective of the civil branch of government appears to be more encompassing and taking into consideration the needs of both men and women in the local populations, the military seems mainly concerned about how its personnel relates to the local populations.

In the first version of the NAP that the United Kingdom government published, gender-based violence was linked to gender justice. It aimed "to promote justice for women and tackle gender-based violence in post-conflict situations."[45] This included measures such as "[t]o advocate for gender-based violence to be included in the mandates of transitional justice mechanisms, including tribunals, reparations, vetting and truth commissions" and "[t]o continue to include tackling gender-based violence and access to justice for women as a priority in DFID-supported Security, Safety and Access to Justice programmes in conflict related countries."[46] As specified in the second version of the NAP, the protection of women is, however, considered to be not only the sole responsibility of DfID, but also that of the FCO.[47] Furthermore, gender training, the protection of women, and institutional changes were assumed to be linked, as one participant of the training for the Female Engagement Team (FET) indicated, stressing "the overall contribution they [the FETs] can make to the protection of women in Afghanistan."[48]

The protection of women from gender-based violence is considered equally important in the Swedish NAP, which states that "[s]pecial attention is to be paid, both in the short term and the long term, to issues concerning the security and protection of women and girls in operational areas," but it also includes "the need for initiatives for sexual and reproductive health and rights."[49] Relevant actions should "build on women's own analysis of the need for protection."[50]

The role and needs of women related to peace agreements

As with respect to the protection of women from gender-based violence, the information provided by the German government on how it supports the role and needs of women in peace processes reflects that the implementation of SCR 1325 is mainly left to those governmental actors active in development cooperation. In Afghanistan, the German government has supported measures to increase the participation of Afghan women in public life, by financing, for example, the training of female judges, attorneys and police officers.[51] Furthermore, the government also supports the participation of women in peace processes in other areas of the world, such as in Latin America.[52] Overall, these projects are administered and implemented by both governmental and nongovernmental agencies that mainly work in development aid.

In the second UK NAP, the role and needs of women related to peace processes are taken account of in three country plans concerning specific actions in Afghanistan, the Democratic Republic of the Congo (DRC) and Nepal. With respect to Afghanistan, for example, the NAP calls on the "Afghan Security Policy and peace-building efforts [… to] include local women's groups" for "[s]ecuring the needs and priorities of women." It also aims to "[i]ncrease the number, influence, and capacity of female Afghans in public life" including members of parliament, ministers and civil servants and to "[i]ncrease female participation in security and justice structures," by, for example, "[f]unding and training to increase and build the capacity of female police officers in Helmand province."[53] Contrary to Germany, the governmental agency spearheading the implementation of the above-named measures is the FCO and only to a lesser extent DfID.

The Swedish NAP also stresses the aim of "[w]omen in conflict areas to participate fully and on equal terms with men at all levels in mechanisms and institutions for conflict prevention, crisis management, peace-building, humanitarian operations, and other initiatives during a post-conflict phase." Therefore, "[n]ational and local women's organisations […] should be supported" and "women's participation in conflict prevention initiatives, conflict management and peace-building" should be increased by "strategic financial contributions and other forms of capacity-building support."[54]

As far as the incorporation of a gender perspective into peacebuilding is concerned, the United Kingdom and Sweden appear to have a more encompassing approach since, in addition to women, men and boys are also acknowledged. With respect to Afghanistan, for example, the United Kingdom plan envisions that the FCO fund a study "to identify how different sections of society, including women, experience 'security' and how they might wish to be protected in future."[55] Evidence for a gender perspective can also be found in references to research commissioned by DfID on "how policies for increased economic growth for women and young men affect conflict and stability in Helmand."[56] In the case of Sweden, the Folke Bernadotte Academy stresses as part of its training that "[a] gender perspective is thus ultimately about not approaching groups of people as homogenous entities, but to see how people's different social positions affect their abilities to access and control resources and take an active part in decisions that affect their lives." It therefore considers it important to "look […] at how the social relations between girls/women and boys/men generate different needs and abilities," and to be aware that "other social structures like class, religion, ethnicity, political belonging, sexual preference or race affect and lead to further differentiation between men as well as between women."[57]

Analysis

Overall, the three reports issued by the German government list a variety of measures that contribute to the implementation of SCR 1325 by German actors. Nevertheless, they do not form a coherent policy regarding the implementation of the resolution. When reading through the reports, the impression one is left with is that the German government listed all gender-related measures, whether they are of relevance for peace and conflict resolution or not. Hence, measures aimed at reconciling family and professional life are mentioned in the same breath as are percentages of female members of parliament and individual measures the government carries out to support women in management positions within Germany.[58] This incoherence might be attributed to, on the one hand, the fact that governmental actors conceive of gender mainstreaming as a cross-section task that should be taken into account and carried out by all governmental agencies at all times, and, on the other hand, the absence of an overarching and integrative document.[59]

In contrast, the implementation processes in the United Kingdom and in Sweden appear to be more coherent. Unlike in Germany, civilian and military governmental actors in the two countries participate as equal stakeholders in the implementation process, are subscribing to a broader agenda, and are working together. One example of this rather open implementation process in Sweden is Gender Force. Founded in 2006 with funding from the EU's Equal Initiative, it is a network comprising governmental and civil society actors, military and civilian actors. While initially intended as a timely confined initiative, Gender Force has meanwhile been firmly integrated into the government structure. The members include the Swedish Armed Forces, the Swedish Police, the Swedish Rescue Service Agency, the NGO Kvinna till Kvinna (Women for Women), the Association of Military Officers in Sweden, and the Swedish Women's Voluntary Defence Organization.[60] Gender Force works, for example, "to improve the Swedish international relief operations and peacekeeping missions by ensuring women's participation."[61] Towards this end, participating members have analyzed recruitment processes and policy documents from a gender perspective with the aim of identifying areas of improvement and many have developed training methods and tools to raise awareness and knowledge of gender issues, in general, and SCR 1325, in particular. Moreover, in 2007 Gender Force published a booklet entitled "Good and Bad Examples: Lessons learned from working with United Nations Resolution 1325 in international missions," which offered concrete examples of how personnel in civilian and military operations had applied SCR 1325.[62]

Compared to the German government, the United Kingdom and Swedish governments' interpretations of SCR 1325 seem more encompassing and to have more transformative potential. Aside from simply paying greater attention to women, the NAPs of the two countries are indicative of fundamental policy and institutional changes. The United Kingdom NAP not only identifies equal treatment as an important strategy to increase the overall representation of women but also lists special measures or gender-specific instruments, such as gender training, to take into consideration both women's and men's interests as well as their needs. Similarly, the Swedish NAP acknowledges that "[w]ork to implement Resolution 1325 must be much more of a shared common concern of all actors and also of both women and men." SCR 1325 is identified as "both an aim in its own right and a way of reaching the objectives of security, development, defence and gender equality policies." While the Swedish NAPs describe the overall goal of gender mainstreaming "to make visible and strengthen women's participation, power, influence, importance, security and enjoyment of their human rights before, during and after conflicts, [… and] conceives of it as a means to bring about democracy, development, and increased respect for human rights,"[63] they promise even more fundamental changes in policies and decision-making processes by calling for, among other things, gender training and new methods of recruitment.

One possible explanation for the observed differences between Germany, on the one hand, and Sweden and the United Kingdom, on the other hand, might be the governments' overall gender mainstreaming policies. It could be argued that Sweden has both the longest experience in regard to public gender policies with a women's policy agency in place since the mid-1970s as well as the most far-reaching gender policies of the three countries aiming at " … incorporat[ing] a gender equality perspective into all policy areas and all activities carried on by the Government Offices."[64] Thus, it does not seem too surprising that it also pursues an encompassing approach when it comes to the implementation of SCR 1325. Similarly, the rather modest and inconsistent response of the German government to the resolution seems to be in line and reflective of the rather late development of gender policies in the 1990s. In the case of the United Kingdom, however, the general gender mainstreaming policies appear less suited to explain the encompassing approach of the government to implement SCR 1325. Like Germany, the British government can be considered a late starter and also only began to adopt relevant gender legislation in the 1990s.

Instead of the overall gender mainstreaming policies, domestic actor constellations seem more telling as to why we find variation across the three countries in the implementation of the Security Council resolution. According to Yvonne Benschop and Mieke Verloo, crucial differences exist between civil society and governmental actors. While the former are expected to favor a broad understanding of gender mainstreaming, the latter, by comparison, have a more narrow and limited conception.[65] According to this line of thinking, the more encompassing understanding of gender mainstreaming in the NAPs of both Sweden and the United Kingdom can be attributed to the participation of civil society actors which was comparatively high, compared to Germany where it was rather low.

In both Sweden and the United Kingdom, governments recognized the importance of civil society in the implementation of SCR 1325. They initiated consultations and shared drafts with representatives of women's organizations and networks. As a result, the respective plans contained not only more far-reaching commitments, but civil society also figures more prominently in them. The impact of civil society organizations is especially noticeable in the United Kingdom where the government responded to criticism of women's organizations in the second version of the NAP, by proposing clear objectives, means of how to accomplish these, and indicators to evaluate the progress of implementation, including a detailed timetable for review.[66] In Germany, by comparison, where civil society organizations were excluded from the implementation process, the dominant interpretation of gender mainstreaming remains a rather narrow one.

In addition to the participation of civil society in the implementation of gender mainstreaming, "[p]ower differences between the parties determine the outcome" according to Benschop and Verloo.[67] In Germany, these existed not only between civil society actors and members of government, but also within the government. Contrary to the ministries of foreign affairs and defense, the governmental agencies concerned with development aid and with a more encompassing understanding of gender mainstreaming only played the role of a junior partner. This stands in stark contrast to the United Kingdom where the FCO, the MoD and DfID participated as "equal stakeholders" in the development of the national action plan.[68] It also differs from Sweden. Although the Ministry of Foreign Affairs took sole responsibility for the implementation of the plan, the views of military and civilian actors were accounted for through the establishment of Gender Force.

Conclusion

In this paper, we examined the implementation of SCR 1325 "Women, Peace, and Security" which calls for gender mainstreaming of peacekeeping and conflict resolution. The case studies of Germany, the United Kingdom, and Sweden illustrate how an international resolution of the UN Security Council is translated, appropriated, and interpreted in different ways on the national level. Understandings regarding gender mainstreaming varied not only across the three countries, but also among actors within them.

Aside from simply paying greater attention to gender, the NAPs of the United Kingdom and Sweden as well as the initiatives and activities already undertaken regarding SCR 1325 are indicative of more fundamental policy and institutional changes. Implementation strategies include equal treatment to increase the overall representation, special measures to take into consideration women's interests as well as needs and gender-specific instruments, such as gender training or impact assessments. In comparison, Germany has as of yet no NAP and governmental policies, with the exception of development aid agencies. The dominant interpretation of gender mainstreaming focused rather narrowly on the internal organizational structure of the military and its relation to the local populations.

These findings are important in several respects: First, they temper the enthusiasm of those who celebrated SCR 1325 as revolutionary. While the adoption of such a resolution can indeed be viewed as an achievement, the results regarding its implementation are, thus far, rather mixed. This insight may have repercussions for UN Women whose mandate pertains not only to developing, but all UN member states. Rather than assuming that the legislation and the policies in "willing" countries are per definition in accordance with an international agreement, the results of this analysis instead suggest, that the organization must also actively work for and ensure compliance in these cases. Second, the empirical evidence gleaned from the various case studies shows that domestic influence does not lessen once a resolution has been agreed upon at the international level. Because the provisions of an international resolution can be interpreted in different ways, new policies, laws, or discursive practices are not always sufficient indicators for the successful implementation of a resolution. As Emanuela Lombardo points out, "what is labeled equal opportunity in one context can be very similar to what is called empowerment in another, while the content of similar labels can diverge tremendously."[69] In the three cases we examined, we found varied interpretations among the involved domestic actors, the civilian and military branches of government, on the one hand, and civil society and governmental actors, on the other hand. Furthermore, the analysis of the implementation of SCR 1325 suggests that the interpretation of gender mainstreaming is more far-reaching and encompassing when civil society actors participate in the domestic implementation process. Finally, studying how international norms are appropriated and translated in comparative perspective is important because it can offer a more realistic assessment of what a government does. It allows us to detect whether what is celebrated as norm acceptance is backed by words and deeds or amounts to little more than rhetoric and affirms the status quo.

Notes

1 *Resolution 1325 (2000)* (Security Council document S/RES/1325), 31 October 2000.

2 Ibid.

3 *Statement by the President of the Security Council* (document S/PRST/2004/40), 28 October 2004; *Statement by the President of the Security Council* (document S/PRST/2005/52), 27 October 2005.

4 Peacewomen 2011, *National Action Plans*, http://www.peacewomen.org/pages/about-1325/national-action-plans-naps.

5 Elissa Helms, "Gender Essentialism and Women's Post-War Activism in Bosnia-Herzegovina," in *Feminists under Fire: Exchanges across War Zones,* ed. Wenona Giles, Malathi de Alwis, Edith Klein, Neluka Silva and Maja Korac (Toronto: Between the Lines, 2003), 181–98; Dyan Mazurana, Angela Raven-Roberts and Jane Parpart, ed., *Gender, Conflict, and Peacekeeping* (Oxford: Rowman and Littlefield, 2005); Johanna Valenius, "A Few Kind of Women: Gender Essentialism and Nordic Peacekeeping Operations," *International Peacekeeping* 14, no. 4 (2007): 510–23; Sandra Whitworth, *Men, Militarism, and Peacekeeping* (Boulder, CO: Lynne Rienner Publishers, 2004).

6 Aisling Swaine, "Assessing the Potential of National Action Plans to Advance Implementation of United Nations Security Council Resolution 1325," *Yearbook of International Humanitarian Law* 12 (2009): 403–33.

7 Belgin F. Gumru and Jan Marie Fritz, "Women, Peace and Security: An Analysis of the National Action Plans Developed in Response to UN Security Council Resolution 1325," *Societies Without Borders* 4, no. 2 (2009): 209–25.

8 Giji Gya, Charlotte Isaksson and Marta Martinelli, *Report on ESDP missions in the Democratic Republic of the Congo (DRC). Final Report.* Paper prepared for UNIFEM Brussels, p. 3, September 2008, revised January 2009, http://www.isis-europe.eu/sites/default/files/programmes-downloads/2009_artrel_242_esdp%26drc-gender-report.pdf.

9 Ray Murphy, "An Assessment of UN efforts to Address Sexual Misconduct by Peacekeeping Personnel," *International Peackeeping* 13, no. 4 (2006): 531–46.

10 Teresa Rees, "Reflections on the Uneven Development of Gender Mainstreaming in Europe," *International Feminist Journal of Politics* 7, no. 4 (2005): 555–74; 560.

11 *Resolution 1325.*

12 UK Government 2006, *UK National Action Plan to Implement UNSCR1325*, http://www.un.org/womenwatch/feature/wps/UK_action_plan_public_version_Sept_06.pdf.

13 UK Government 2010, *UK Government National Action Plan on UNSCR 1325 Women, Peace & Security*, http://www.fco.gov.uk/resources/en/pdf/global-issues/conflict-prevention/women-peace-security-action-plan.

14 Swedish Government 2006, *The Swedish Government's Action Plan to Implement Security Council Resolution 1325 (2000) on Women, Peace and Security*, HDIM.DEL/92/06, 4 October, http://www.un-instraw.org/en/gps/gps-homepage/national-action-plans-on-resolution-1325-8.html#Sweden.

15 Swedish Government 2009, *The Swedish Government's Action Plan for 2009–2012 to Implement Security Council Resolution 1325 (2000) on Women, Peace and Security*, http://www.un.org/womenwatch/ianwge/taskforces/wps/nap/Swedish_Action_Plan_final_version.pdf.

16 Bundesregierung, *Bericht der Bundesregierung an den Generalsekretär der Vereinten Nationen zur Umsetzung von SR-Resolution 1325*. Stand 18 June 2004 (Berlin, 2004); Bundesregierung, *Bericht der Bundesregierung über Maßnahmen zur Umsetzung der Sicherheitsratsresolution 1325 (2000) "Frauen, Frieden und Sicherheit"*, 19 November 2007 (Berlin, 2007); Bundesregierung, *Bericht der Bundesregierung über Maßnahmen zur Umsetzung der Sicherheitsratsresolution 1325 ("Frauen, Frieden und Sicherheit")*, November 2010 (Berlin, 2010).

17 BMVg, *Stellungnahme des Bundesministeriums der Verteidigung zum Thema "Gleichberechtigung und Akzeptanz von Frauen in der Bundeswehr,"* Berlin, 4 March 2009.

18 Bundesregierung 2010, *Bericht der Bundesregierung über Maßnahmen zur Umsetzung der Sicherheitsratsresolution 1325*, 16.

19 Ministry of Defence, *Women in the Armed Forces*, http://www.mod.uk/DefenceInternet/FactSheets/PersonnelFactsheets/WomenInTheArmedForces.htm.

20 Anita Schjølset 2010, *NATO and the Women. Exploring the Gender Gap in the Armed Forces,* Peace Research Institute Oslo (PRIO), http://www.prio.no/sptrans/58538131/Schjolset_FDreport_July2010_final_.pdf, 31.

21 Feministisches Institut 2006, *Frieden und Sicherheit für alle. Eine feministische Kritik der gegenwärtigen Sicherheitspolitik.* Diskussionspapier erstellt vom Feministischen Institut der Heinrich Böll Stiftung auf Basis der fachlichen Arbeit der AG "Gender in der Friedens-Sicherheitspolitik und Krisenprävention," 53, June, http://www.glow-boell.de/media/de/txt_rubrik_2/Positionspapier_2006_deutsch.pdf.

22 Bundesregierung 2007, *Bericht der Bundesregierung über Maßnahmen zur Umsetzung der Sicherheitsratsresolution 1325 (2000) "Frauen, Frieden und Sicherheit,"* 9.

23 UK Government 2006, *UK National Action Plan to Implement UNSCR 1325*, 8.

24 Swedish Government 2006, *The Swedish Government's Action Plan to Implement Security Council Resolution 1325 (2000) on Women, Peace and Security*, 6, 17, 17; see also Swedish Government 2009, *The Swedish Government's Action Plan for 2009–2012 to Implement Security Council Resolution 1325 (2000) on Women, Peace and Security*, 12.
25 Swedish Government 2009, *The Swedish Government's Action Plan for 2009–2012 to Implement Security Council Resolution 1325 (2000) on Women, Peace and Security*, 11, 12.
26 Bundeswehr 2009, *Starke Truppe—Immer mehr Frauen entscheiden sich für die Bundeswehr*, http://www.bundeswehr.de; Bundesregierung 2010, *Bericht der Bundesregierung über Maßnahmen zur Umsetzung der Sicherheitsratsresolution 1325 ("Frauen, Frieden und Sicherheit")*, 15.
27 Bundesregierung 2007, *Bericht der Bundesregierung über Maßnahmen zur Umsetzung der Sicherheitsratsresolution 1325 (2000) "Frauen, Frieden und Sicherheit,"* 9.
28 Bundesregierung 2004, *Bericht der Bundesregierung an den Generalsekretär der Vereinten Nationen zur Umsetzung von SR-Resolution* 1325, 14; Bundesregierung 2007, *Bericht der Bundesregierung über Maßnahmen zur Umsetzung der Sicherheitsratsresolution 1325 (2000) "Frauen, Frieden und Sicherheit,"* 9.
29 Bundesregierung 2010, *Bericht der Bundesregierung über Maßnahmen zur Umsetzung der Sicherheitsratsresolution 1325 ("Frauen, Frieden und Sicherheit")*, 16.
30 UK Government 2010, *UK Government National Action Plan on UNSCR 1325 Women, Peace & Security*, 11.
31 Ibid., 11–12.
32 Ministry of Defence, 2011, *Female Engagement Teams Support Afghan Women*, http://www.mod.uk/DefenceInternet/DefenceNews/MilitaryOperations/FemaleEngagementTeamsSupportAfghanWomen.htm.
33 Stabilisation Unit 2011, *Female Engagement Teams Prepare for Deployment to Afghanistan*, http://www.stabilisationunit.gov.uk/newsroom/534-female-engagement-teams-prepare-for-deployment-to-afghanistan.html.
34 Swedish Government 2006, *The Swedish Government's Action Plan to Implement Security Council Resolution 1325 (2000) on Women, Peace and Security*, 17; Swedish Government 2009, *The Swedish Government's Action Plan for 2009–2012 to Implement Security Council Resolution 1325 (2000) on Women, Peace and Security*, 13.
35 Folke Bernadotte Academy 2009, *Gender Force*, http://www.folkebernadotteacademy.se/en/Training/FBA-and-UNSCR-1325/Gender-Force-/.
36 Folke Bernadotte Academy 2011, *Gender Field Adviser*, http://www.folkebernadotteacademy.se/en/Training/Courses1/Gender-Field-Advisor/.
37 Folke Bernadotte Adacemy 2009, *The Folke Bernadotte Academy*, http://www.folkebernadotteacademy.se/en/About-FBA/.
38 Ibid.
39 Folke Bernadotte Academy 2010, *Leadership and Gender in Peace Operations*, http://www.folkebernadotteacademy.se/en/Training/Courses1/Leadership-and-Gender-in-Peace-Operations/.
40 Folke Bernadotte Academy 2009, *The Folke Bernadotte Academy and Project UN Security Council Resolution 1325*, http://www.folkebernadotteacademy.se/en/Training/FBA-and-UNSCR-1325/.
41 Ibid.

42 Bundesregierung 2007, *Bericht der Bundesregierung über Maßnahmen zur Umsetzung der Sicherheitsratsresolution 1325 (2000) "Frauen, Frieden und Sicherheit,"* 14.
43 GTZ 2006, *Gender und Konflikte. Ein Orientierungspapier*, http://www2.gtz.de/dokumente/bib/07–0144.pdf
44 Bundesregierung 2007, *Bericht der Bundesregierung über Maßnahmen zur Umsetzung der Sicherheitsratsresolution 1325 (2000) "Frauen, Frieden und Sicherheit,"* 15.
45 UK Government 2006, *UK National Action Plan to Implement UNSCR1325*, 10.
46 Ibid., 10.
47 Ibid.
48 Stabilisation Unit 2011, *Female Engagement Teams Prepare for Deployment to Afghanistan.*
49 Swedish Government 2009, *The Swedish Government's Action Plan for 2009–2012 to Implement Security Council Resolution 1325 (2000) on Women, Peace and Security*, 13.
50 Swedish Government 2006, *The Swedish Government's Action Plan to Implement Security Council Resolution 1325 (2000) on Women, Peace and Security*, 10.
51 Bundesregierung 2007, *Bericht der Bundesregierung über Maßnahmen zur Umsetzung der Sicherheitsratsresolution 1325 (2000) "Frauen, Frieden und Sicherheit,"* 25.
52 Ibid., 28.
53 UK Government 2010, *UK Government National Action Plan on UNSCR 1325 Women, Peace & Security*, 18, 19, 19–20, 22, 22.
54 Swedish Government 2009, *The Swedish Government's Action Plan for 2009–2012 to Implement Security Council Resolution 1325 (2000) on Women, Peace and Security*, 9, 14, 14, 14.
55 UK Government 2010, *UK Government National Action Plan on UNSCR 1325 Women, Peace & Security*, 19.
56 Ibid., 25.
57 Folke Bernadotte Academy 2009, *The Folke Bernadotte Academy and Project UN Security Council Resolution 1325*, http://www.folkebernadotteacademy.se/en/Competences/UN-resolution-1325/.
58 Bundesregierung 2004, *Bericht der Bundesregierung an den Generalsekretär der Vereinten Nationen zur Umsetzung von SR-Resolution 1325*, 4
59 Ibid., 3–5; Bundesregierung 2007, *Bericht der Bundesregierung über Maßnahmen zur Umsetzung der Sicherheitsratsresolution 1325 (2000) "Frauen, Frieden und Sicherheit,"* 5–6; Bundeswehr, *NATO-Komitee—Potenziale nutzen* (Berlin 2007).
60 Swedish Government 2009, *The Swedish Government's Action Plan for 2009–2012 to Implement Security Council Resolution 1325 (2000) on Women, Peace and Security*, 13.
61 Gender Force Sweden 2010, *About Gender Force Sweden*, http://www.genderforce.se/About_Genderforce.htm.
62 Gender Force Sweden 2007, *Good and Bad Examples: Lessons Learned from Working with United Nations Resolution 1325 in International Missions*, http://www.genderforce.se/dokument/Good_and_bad_examples_English_A4.pdf.

63 Swedish Government 2006, *The Swedish Government's Action Plan to Implement Security Council Resolution 1325 (2000) on Women, Peace and Security*, 11, 2, 2.

64 Ibid., 2.

65 Yvonne Benschop and Mieke Verloo, "Sisyphus' Sisters: Can Gender Mainstreaming Escape the Genderedness of Organizations?" *Journal of Gender Studies* 15, no. 1 (2006): 19–33 (31).

66 UK Government 2010, *UK Government National Action Plan on UNSCR 1325 Women, Peace & Security.*

67 Benschop and Verloo, "Sisyphus' Sisters: Can Gender Mainstreaming Escape the Genderedness of Organizations?" 19–33 (31).

68 UK Government 2006, *UK National Action Plan to Implement UNSCR1325*, 1.

69 Emanuela Lombardo, "Integrating or Setting the Agenda? Gender Mainstreaming in the European Constitution-Making Process," *Social Politics: International Studies in Gender, State & Society*, special issue "Gender Mainstreaming" 12, no. 3 (2005): 412–32.

11 How do women's rights norms travel?

Peace-building operations as opportunity structures in Bosnia

Anne Jenichen

- **International peace-building, security and women's rights**
- **The institutionalization of women's rights in Bosnia and Herzegovina**
- **Conclusions**

The ethno-nationalist war in Bosnia and Herzegovina (BiH) between 1992 and 1995 became infamous for the massive use of sexualized violence against women.[1] The war in Bosnia, in fact, was the case which brought sexualized violence in war situations onto the international agenda, eventually leading to its international recognition as war crime and crime against humanity.[2] Despite international attention to the issue, the international post-war intervention in the country started out as a completely gender-blind endeavor. This is furthermore striking because the Fourth World Conference on Women in Beijing had taken place just a few months before the peace agreement and the international mandate in Bosnia were hammered out. The final document of this conference, the Beijing Platform for Action, refers to the protection of women in violent conflicts and the promotion of the participation of women in peace negotiations and post-war politics as one of its 12 strategic goals.[3] Even though delegations from all states involved in the post-war intervention in Bosnia had also participated in the Beijing Conference, women's rights issues were utterly ignored by the intervening international community—with detrimental effects for the security and rights of women.[4]

Bosnia is a telling example of the neglect of gender issues in international post-war reconstruction. However, it is not the only one. Given the still insufficient inclusion of gender issues in peace processes, it may come as a surprise that women's rights norms travel to post-war states nonetheless. If not the intervening international community, who else initiates these processes? Which role exactly do international peace-building operations play in the diffusion of women's rights norms to post-war contexts? How can domestic feminist agents benefit from post-war interventions? These specific questions have thus far been neglected by research on post-war reconstruction, on women's rights in these processes as well as on the global diffusion of women's rights. I argue in this chapter that it is usually not the intervening international community that imposes international women's rights norms in these contexts. Rather, domestic "norm entrepreneurs"[5] advocate the institutionalization of international women's rights norms, using international peace-building operations as specific opportunity structures.[6] International peace-building operations often create a specific institutional context which, in turn, affects the strategies of domestic norm entrepreneurs. By providing potential allies and specific resources and constituting an internationalized agenda, they increase the opportunities for domestic norm entrepreneurs to convince domestic policy-makers of the necessity of their women's rights demands.

The reconstruction of the norm diffusion mechanism presented here is primarily based on in-depth field research in post-war Bosnia and Herzegovina. Yet, there is evidence from other post-war contexts subject to an international post-war intervention which largely confirms the key findings identified here.[7]

The chapter starts with some insights from the debate on women's rights norms in the governance of international security. It proceeds with the case study on Bosnia and Herzegovina, illustrating how domestic norm entrepreneurs set the agenda for innovative women's rights policy change in the country and how they made use of the opportunities offered by the international peace-building operation. The conclusion reflects more in general on how international peace-building operations serve as specific opportunity structures in the diffusion of women's rights norms and policies after wars.

International peace-building, security and women's rights

International peace-building has become a major issue in global governance: Former United Nations (UN) Secretary-General Boutros Boutros-Ghali (1991–96) identified "post-conflict peace-building" as one of the UN's major tasks in addressing violent conflicts. In his "An Agenda for Peace," he defined the concept as "... action to identify and support structures which will tend to strengthen and solidify peace in order to avoid a relapse into conflict. ... [P]ost-conflict peace-building ... can prevent the recurrence of violence among nations and peoples."[8]

Wars and violent conflicts usually implicate a breakdown of social and political institutions. Post-war situations, thus, often come along with weak political and administrative institutions lacking the capacities to enforce political decisions. International and regional governmental organizations, such as the UN or the Organization for Security and Co-operation in Europe (OSCE), therefore send international peacebuilding operations[9] to support the post-war state (or state-like entity) in its endeavor to build peace and to further democracy and development. The overall task of most international post-war interventions is to oversee the implementation of a peace agreement. However, their mandates nowadays do not refer to issues of security alone but extend to a variety of civil issues, such as good governance, rule of law, human rights as well as development and humanitarian aid. Increasingly, the promotion of liberal democracy and economic reforms has received a central position within their mandates.[10] In some—rarer—cases, international organizations temporarily assume governmental and administrative functions to ensure the transition to a stable and democratic post-war order.[11] In these contexts, international transitional administrations possess extensive legislative and executive authorities to govern domestic policy processes independently. But even if international peacekeeping operations lack such external authorities,[12] they contribute to the internationalization of the post-war agenda by promoting compliance with international norms, by contributing to their specification and implementation, as well as by their overall monitoring and advice. The heavy presence and influence of international advisors and policy-makers in post-war contexts significantly affect domestic policy-making processes which, therefore, cannot be understood properly without taking the interaction between domestic and international actors into account.[13]

From a feminist perspective, multidimensional mandates promoting post-war security, democracy and human rights would also have to include the advancement of women's rights. Not only *are* women's rights human rights, as internationally endorsed at the World Conference on Human Rights 1993 in Vienna,[14] but women's equal participation in political decision-making is also an important part of democracy, if we take the concept seriously.[15] The feminist debate on security, finally, has worked towards moving the concept of security away from its initially narrow focus on military security towards a broader, multidimensional concept including individual security of men *and* women as well as, beyond physical security, also structural security in terms of economic and social justice.[16] Thus, "[a]ny feminist perspective would argue that a truly comprehensive security cannot be achieved until gender relations of domination and subordination are eliminated."[17]

Beyond these reasons to include the promotion of women and their rights in peace-building efforts, there exist a number of international norms demanding respect for women's rights in post-war contexts. The most prominent one certainly is UN Security Council Resolution 1325 on women, peace and security.[18] This resolution instructs UN member states and the UN itself to ensure equal participation of women and consider women's specific needs and interests in peace processes. Its adoption has been a milestone regarding the integration of women's rights into the global governance of security, even though its implementation is progressing extremely slowly.[19] Furthermore, there is the international Convention on the Elimination of All Forms of Discrimination against Women (CEDAW), prohibiting any form of discrimination against women in general, both in the public and the private sphere. CEDAW was adopted in 1979 and entered into force in 1981. It has been ratified by 186 states so far, including states torn by political violence and war. In addition, the declarations and platforms for action of the international women's conferences in Mexico City (1975), Copenhagen (1980), Nairobi (1985), and especially Beijing (1995), though not legally binding, provide further important guidelines for the formulation of policy programs and the engineering of institutions. Finally, there are also regional organizations, such as the European Union (EU), defining gender equality norms which are legally binding for member states of these organizations.

International peace-building operations, however, long ignored women's rights issues nearly completely. This changed with the adoption of UN Security Council Resolution 1325 in 2000, which led to the integration of gender advisors and units into UN peacekeeping operations and of gender issues into their mandates. But even after its adoption, these gender entities have often been criticized for being under-resourced and largely ineffective.[20] Their activities, furthermore, usually focus on providing training and organizing projects on issues such as interethnic dialogue or income generation for women, rather than on promoting the adoption and implementation of domestic women's rights norms and policies directly.[21]

Two questions arise from this observation: First, why have international peace-building operations largely refrained from pushing women's rights *policy* issues? It could be argued in response that in most cases, these decisions were based on lacking political will within the international peace-building operation, rather than on deliberate decisions reflecting acknowledgement of the greater legitimacy of domestically driven policy processes. Research on gender issues in internationally promoted peace processes suggests this answer by highlighting the lacking resolve of international, as well as national, elites to acknowledge (or recognize) the relevance of women's rights issues in peace processes.[22] The second puzzle then is to explain why women's rights norms and policies have spread to post-war contexts nevertheless. The following section provides an answer to this question using the diffusion of women's rights norms into the special context of Bosnia and Herzegovina as an example.

The institutionalization of women's rights in Bosnia and Herzegovina

Until the war of secession, Bosnia and Herzegovina (BiH) had been one of the eight republics of the Socialist Federal Republic of Yugoslavia, which, since its inception, declared the elimination of inequalities between women and men official state policy. After World War II, women received legal equality with men in all areas of state, economy and social life. Mothers and children were guaranteed particular protection. Equality was primarily promoted in the fields of education and employment. In other fields, however, the promotion was limited to public declarations by the Yugoslav government, with no active policies following. Women, consequently, despite legal equality and increasing participation in education and employment, remained disadvantaged in the former Yugoslavia. They got positions with less pay and less prestige than men's, enjoyed less job security, worse working conditions, and less decision-making power. In leadership positions, women remained underrepresented.[23] Other fields pertaining to women's rights were completely ignored, such as combating gender-specific stereotypes through the education system,[24] or promoting gender equality and tackling violence against women in the private sphere. Overall, it was primarily the area of marriage and family which turned out to be considerably resistant to changes.[25]

Consequently, there had been legislation, as well as some independent organizing of women,[26] which norm entrepreneurs after the war could—and did—build upon. However, the mentioned deficits made improvements necessary, particularly in the areas of violence against women and gender-based discrimination. The war had left women in precarious situations, with their almost complete exclusion from political decision-making, discrimination on the labor market, and high amounts of violence against them. Moreover, nationalist discourses and politics accompanying the ethno-political conflict in BiH, the association of gender equality with the discredited communist regime, and prevailing gender-based stereotypes have constituted major barriers to the advancement of the situation and rights of women. Until today, political elites do not display the political will and do not allocate the resources necessary to successfully combat gender inequalities. The social status of women, consequently, has not changed substantially since the war ended, despite the increasing formal acceptance of women's rights norms. To date, gender relations in BiH have been characterized by political, economic, and social inequality. Women are politically under-represented, have less access to economic resources, and gender-based violence is widespread, to name just a few of the current problems.[27]

Yet, despite the unfavorable starting position, women's rights norms were broadly institutionalized in BiH after the war. Between 1998 and 2006, about 20 new laws, by-laws and action plans were adopted in BiH to promote women's rights and gender equality.[28] Even though implementation is insufficient, it is still relevant to ask why this legal framework was introduced in the first place, given that national political elites as well as the majority of the intervening international community were hardly interested in women's rights issues at that time. Who advocated these policy changes? And how could relevant decision-makers be convinced of their necessity?

The institutionalization of women's rights norms in BiH neither was imposed by the intervening international community nor did it emerge as a result of domestic norm entrepreneurs' activities alone. It rather resulted from the cooperation of domestic women's rights activists and politicians, on the one hand, with a few members of the international community, on the other. However, the institutionalization of women's rights norms was domestically driven in the first place. Women from Bosnian women's nongovernmental organizations (NGOs) and political parties initiated these reform processes by putting them on the public agenda. But they had not been able to persuade domestic policy-makers of the necessity to formulate and adopt respective policies before they received support from some international representatives.

Bosnian women's rights organizations were the first after the war who attempted to draw attention to the precarious situation of women. Already during the war, they had been committed to combating sexualized (war) violence and to helping refugees and displaced women.[29] After the war, they organized women's conferences,[30] public debates and roundtable discussions. They published and disseminated women's rights reports, such as a first CEDAW Shadow Report in 1999, and tried to mobilize the media. However, the political elite ignored their demands completely, just like the majority of the intervening international community. Neither regarded women's rights issues as relevant for the peace process at that time.

One of the main problems of Bosnian women's rights activists and politicians since the end of the war has been the lack of access to official policy and institutional processes required to convince domestic policy-makers of the necessity of women's rights policies. Moreover, they have been short of time and material resources needed to pursue political campaigns advocating women's rights policies. Most of the time and funding available to women's NGOs is dedicated to the implementation of projects funded by external donors. The agendas of donors, however, seldom focus on political advocacy but usually on issues of practical help to women, such as psycho-social and legal counseling and support as well as trainings and income generation for women.[31] Also women from political parties advocating women's rights policies in BiH have been relatively rare, and the few of their kind lack the necessary support within their own parties—although, important to mention, they come from the whole political spectrum ranging from nationalist to moderate. Domestic state institutions dedicated to the promotion of gender equality, established in BiH in 2000, 2002 and 2005, lack the necessary resources to instigate meaningful change on their own.[32] Given these unfavorable circumstances, Bosnian norm entrepreneurs first turned to international allies in search of support.

During the first decade of the international intervention, there were a few representatives holding leading positions within international organizations, such as the head of the Democratization Department of the Mission of the OSCE in BiH as well as the head of the Office of the High Commissioner for Human Rights (OHCHR) in BiH, who decided to support the endeavors of Bosnian women's rights activists and politicians. These allies also advocated the integration of women's rights and gender equality issues into the agenda of the wider international community in BiH—with mixed success. Although some issues, most prominently trafficking in human beings, gradually received more attention, and international activities in BiH in the field of women's rights and gender equality increased temporarily, the integration was neither systematic nor sustainable.[33] The support of domestic norm entrepreneurs was still left to a few international representatives.

Particularly in the case of the electoral gender quota, introduced in 1998 into the provisional election rules and regulations and transferred in 2001 to the first national election law, the allies from international organizations were crucial in getting access to relevant decision-makers and formal policy processes. They assisted, for instance, in getting access to the then head of the OSCE Mission in BiH in 1998, who, eventually, put the introduction of the electoral gender quota on the official agenda of the Provisional Election Commission. They also facilitated access to the international head of the working group responsible for the drafting of the first national election law and, thus, enabled female Bosnian politicians to convince him of maintaining the gender quota in the new law.[34] The Office of the High Representative (OHR)[35] managed the publication of the previously secret intention of the parliamentary working group, which was responsible for the amendment of the election law in 2005, to reformulate the gender quota to the disadvantage of women, thus enabling domestic women's rights activists, politicians and the governmental gender equality institutions jointly to prevent this plan. In the case of the 2003 Gender Equality Law, the OHCHR and the international NGO "International Human Rights Law Group" assisted Bosnian women's rights activists in monitoring as well as directly contributing to the drafting of the Gender Equality Law before its adoption in 2003.[36] The same international NGO (now under its new name "Global Rights") also supported a Bosnian NGO campaign in 2005 advocating the removal of an article from the draft law on the protection from violence within the family, which was saying that domestic violence, under certain circumstances, could also be dealt with as a misdemeanor only.

The cooperation between domestic and international norm entrepreneurs was characterized by an exchange of different resources. While the international allies provided access, time, expertise on international norms and standards, as well as experiences in legislative reform and legal advocacy, their domestic counterparts contributed their local knowledge and visible support. These unequal resource endowments often led to a specific division of labor, particularly during policy formulation and formal adoption processes. While the international partners primarily assumed the role of coordinating campaigns and carrying out the immediate lobbying within parliamentary institutions, such as working groups drafting legislation and the parliamentary gender commission, their domestic nongovernmental partners mainly supported these activities by lending them their publicly visible support. Representing domestic voices, they thus provided their international partners with the credibility needed to legitimately intervene into domestic policy-making processes.

Two prominent examples of this pattern were the NGO campaigns around the Bosnian gender equality law and the laws on the protection from violence within the family. In the case of the campaign for the gender equality law, a member of the international NGO "International Human Rights Law Group" was sent by its local partners to directly participate in the working group drafting the law.[37] In both campaigns it again was this international NGO which organized public hearings in the parliamentary gender commissions at national and sub-national level to persuade their members to support the draft laws in parliament. Bosnian women's NGOs in the meantime initiated NGO campaigns publicly supporting the formulation and adoption of both laws, which were supported by other human rights NGOs and some academics as well. In later campaigns, such as around the already mentioned amendment of the election law in 2005, the coordinating roles were increasingly assumed by domestic governmental actors, such as the state-level Gender Agency, founded in 2005, as well as by single Bosnian NGOs and female politicians.

These coalitions between domestic and international norm entrepreneurs in post-war states usually exist only temporarily. In the course of the international intervention's gradual retreat from the post-war state, and the simultaneous growth of capacities of domestic state institutions, domestic actors increasingly replace international representatives as crucial allies. In BiH, particularly the governmental institutions for the promotion of gender equality (Gender Centers and Gender Agency, established between 2000 and 2005) have assumed the role of coordinating women's rights policy reform campaigns and of helping women's interests to access the policy process. There is evidence that already since their foundation, they have tried to distance international representatives from domestic women's rights campaigns to increase their own legitimacy in the new policy field.

With the ongoing withdrawal of the international community, domestic norm entrepreneurs will also rely more and more on transnational contacts and cooperation, because domestic norm entrepreneurs are usually more effective if they can mobilize support by allies both within and outside their state, i.e., if they build so called "insider-outsider coalitions."[38] Due to the presence of international allies within the country, transnational mobilizing had not been that necessary for domestic NGOs before. Given the scarcity of resources, it may become a challenging task in the future.

Besides international resources, international post-war interventions also provide opportunities for framing demands in a new way. Women's rights activists in post-war states such as BiH often frame their requests as international obligations of their governments, rather than only as necessary for the realization of the rights of women. They invoke CEDAW, and since 2000 also UN Security Council Resolution 1325, to legitimize their demands, and use the Beijing Platform for Action of 1995 to formulate goals and programs for the promotion of women's rights.[39] In BiH, women's rights norm entrepreneurs additionally framed many of their political demands as European obligations of the Bosnian state.

However, the EU had not insisted on women's rights policy change in BiH itself. They, thus, took not only advantage of the great interest of the Bosnian political elite and general population in the integration of the country into the community of European states,[40] but also of the lacking knowledge of domestic decision-makers about the concrete conditions for accession.[41]

Conclusions

The example of the institutionalization of women's rights in post-war BiH reveals that the internationalized institutional context of a post-war state can represent a specific opportunity structure for domestic norm entrepreneurs by providing potential allies, access to international resources and relevant policy-makers, as well as strong arguments.

In the course of international post-war interventions, there do not only emerge new constraints for domestic agents of change, such as their exclusion from policy-making processes, but also new opportunities. The presence of international peace-building operations and the internationalized agenda provide a specific opportunity structure facilitating international norm diffusion as well as helping domestic norm entrepreneurs to push their own agendas. The presence of international peace-building operations within the post-war state decisively enlarges the spectrum of stakeholders and strategies and makes transnational mobilization of support by domestic norm entrepreneurs less necessary. The geographical proximity, and often even cultural proximity if domestic citizens work in international organizations, makes it easier for domestic norm entrepreneurs to access international resources. They first approach the intervening international community to make their voice heard if domestic policy-makers do not respond to their demands, but also because they perceive the international community in a post-war context as more powerful than their own state institutions.[42] International peace-building operations on their own, however, have not been very active in translating international women's rights norms to local contexts. In BiH, as well as in many other post-war contexts, domestic norm entrepreneurs had to initiate this process themselves.[43] Arguably, a comprehensive understanding of and commitment to gender equality norms in peace-building operations is an institutional responsibility of the UN and the sending states. If UN Women is to implement this responsibility, it should seriously invest in comprehensive gender-awareness in all UN Department of Peacekeeping Operations (DPKO) activities. This would enhance—and should not monopolize—the work of domestic gender equality norm entrepreneurs.

The specific opportunity structure constituted by international post-war interventions also affects the framing strategies of domestic norm entrepreneurs by internationalizing the post-war state's (or state-like entity's) agenda. Intervening international organizations oblige the target state to observe international norms and push for its integration into international structures. Irrespective of whether the target state acknowledges this agenda in principle or not, it must behave accordingly if it wants to retrieve full sovereignty and be internationally recognized as an independent and stable state (at least if the intervening international community sticks to its own standards). Most post-war states, in addition, are interested in humanitarian and development aid as well as international economic cooperation. On that account, recognition and reputation as a reliable partner is needed which can be achieved by meeting international expectations and complying with international norms. Last but not least, internationalized framing strategies also help to attract international support as they address the main responsibility of international organizations: promoting the compliance with international norms.

As argued above, women's security, both in physical and socio-economic terms, should be an important element in peace processes. Promoting the diffusion of women's rights norms to post-war contexts could therefore be an important building block of more comprehensive peace strategies. This chapter demonstrated how such a process could be governed in a multilevel, participatory way. Relying on the agenda-setting of domestic norm entrepreneurs and their support by international allies, in my view, is a more promising and legitimate approach than imposing policy change from above. Whether it can serve as a paradigm for successful norm diffusion processes in post-war contexts in the future remains to be seen.

Notes

1 Tanja Djurić-Kuzmanović, Rada Drezgić, and Dubravka Žarkov, "Gendered War, Gendered Peace: Violent Conflicts in the Balkans and Their Consequences," in *Gendered Peace: Women's Struggles for Post-War Justice and Reconciliation*, ed. Donna Pankhurst (New York: Routledge, 2008), 265–91.

2 Valerie Oosterveld, "Prosecution of Gender-Based Crimes in International Law," in *Gender, Conflict, and Peacekeeping*, ed. Dyan E. Mazurana, Angela Raven-Roberts, and Jane Parpart (Lanham, MD: Rowman & Littlefield, 2005), 67–82.

3 Beijing Platform for Action, Fourth World Conference on Women, Beijing, September 1995.

4 Christine Chinkin and Kate Paradine, "Vision and Reality: Democracy and Citizenship of Women in the Dayton Peace Accords," *The Yale Journal of*

International Law 26, no. 1 (2001): 103–78; Kvinna till Kvinna, *Engendering the Peace Process: A Gender Approach to Dayton—and Beyond* (Stockholm, Sweden: Kvinna till Kvinna, 2000); Madeleine Rees, "International Intervention in Bosnia-Herzegovina: the Cost of Ignoring Gender," in *The Postwar Moment: Militaries, Masculinities and International Peacekeeping*, ed. Cynthia Cockburn and Dubravka Zarkov (London: Lawrence & Wishart, 2002), 51–67.

5 Martha Finnemore and Kathryn Sikkink, "International Norm Dynamics and Political Change," *International Organization* 52, no. 4 (1998): 887–917.

6 The term "opportunity structure" stems from social movement and policy research. It describes scope conditions which determine or at least strongly affect the opportunities of actors to articulate their protest, demand political action, receive attention, and mobilize political support. Opportunity structures do not represent political structures alone but comprise all structures, agents and events within a political system, which are relevant to the mobilization of protest. See Jack A. Goldstone, "More Social Movements or Fewer? Beyond Political Opportunity Structures to Relational Fields," *Theory and Society* 33, no. 3–4 (2004): 333–65.

7 I have compared the case of BiH with similar women's rights policy processes in other post-war/violence contexts, such as Afghanistan, Cambodia, East Timor, Iraq, Kosovo, and Mozambique. The results suggest patterns very similar to the ones in BiH. See Anne Jenichen, "Chapter 7" in *Mehrebenen-Advocacy-Koalitionen in internationalisierten Nachkriegsstaaten: Frauenrechtspolitischer Wandel in Bosnien und Herzegowina*, PhD diss., University of Bremen, 2010.

8 *An Agenda for Peace: Preventive Diplomacy, Peacemaking and Peace-keeping, Report of the Secretary-General Pursuant to the Statement Adopted by the Summit Meeting of the Security Council on 31 January 1992* (Security Council document A/47/277–S/24111), 17 June 1992.

9 The official term used by the UN is "peacekeeping operation," the OSCE term is "field mission." I use "peace-building operation" as an umbrella term here to highlight the multidimensional tasks of these operations as well as the focus on post-war situations.

10 Roland Paris, *At War's End: Building Peace After Civil Conflict* (Cambridge: Cambridge University Press, 2004).

11 Examples are the UN Transitional Authority in Cambodia (UNTAC, 1992–93), the Office of the High Representative in BiH (OHR, since 1997), the UN Interim Administration Mission in Kosovo (UNMIK, 1999–2008), or the UN Transitional Administration in East Timor (UNTAET, 1999–2002).

12 One prominent example is the UN Assistance Mission in Afghanistan (UNAMA). Since this kind of intervention violates the principle of sovereignty the least and, furthermore, costs less than running a transitional administration, it is the most common form of international post-war intervention. Beyond Afghanistan, such interventions also exist in Angola, El Salvador, Georgia, Guatemala, Haiti, the Democratic Republic of Congo, Namibia, Rwanda, Sierra Leone, Tajikistan, and Cyprus. Picking up on the increasing criticism of the instrument of transitional administration, the first Special Representative of the UN Secretary-General for Afghanistan, Lakhdar Brahimi, coined the term "light footprint approach" for these contexts as, in his view, the international community should avoid creating parallel institutions and systems which only undermine local authority,

hamper coordination and cause competition. See Lakhdar Brahimi, "State Building in Crisis and Post-Conflict Countries," paper presented at 7th Global Forum on Reinventing Government, Building Trust in Government, Vienna, 26–29 June 2007, http://unpan1.un.org/intradoc/groups/public/documents/un/unpan026305.pdf.

13 Similarly argued by Julie Ballington and Drude Dahlerup, "Gender Quotas in Post-conflict States: East Timor, Afghanistan and Iraq," in *Women, Quotas and Politics*, ed. Drude Dahlerup (Abingdon, UK: Routledge, 2006), 249–58; Louise Olsson, *Gender Equality and United Nations Peace Operations in Timor Leste* (Leiden, The Netherlands: Martinus Nijhoff Publishers, 2009); Daniella Christova Schmitt, *Do Transitional Administrations Fail? A Comparative Study of the Kosovo and East Timor Experiences* (Saarbrücken, Germany: Südwestdeutscher Verlag für Hochschulschriften, 2009); Oisín Tansey, *Regime-Building: Democratization and International Administration* (Oxford: Oxford University Press, 2009).

14 Jutta M. Joachim, *Agenda Setting, The UN, and NGOs: Gender Violence and Reproductive Rights* (Washington, DC: Georgetown University Press, 2007).

15 Anne Phillips, *Engendering Democracy* (Cambridge: Polity Press, 1991).

16 Eric M. Blanchard, "Gender, International Relations, and the Development of Feminist Security Theory," *Signs* 28, no. 4 (2003): 1289–1312; J. Ann Tickner, *Gender in International Relations: Feminist Perspectives on Achieving Global Security* (New York: Columbia University Press, 1992).

17 Tickner, *Gender in International Relations*, 23.

18 *Resolution 1325(2000)* (Security Council document S/RES/1325), 31 October 2000.

19 Christine Bell and Catherine O'Rourke, "Peace Agreements or Pieces of Paper? The Impact of UNSC Resolution 1325 on Peace Processes and their Agreements," *International and Comparative Law Quarterly* 59, no. 4 (2010): 941–80; Susan Willett, "Introduction: Security Council Resolution 1325: Assessing the Impact on Women, Peace and Security," *International Peacekeeping* 17, no. 2 (2010): 142–58. Telling evidence for the lacking implementation is the UN Security Council's decision in 2009 to adopt Resolution 1889 promoting the implementation of 1325.

20 Ana Villellas Ariño and Gema Redondo de la Morena, *An Approach to the Kosovo Post-War Rehabilitation Process from a Gender Perspective* (Barcelona, Spain: Quaderns de Construcció de Pau No. 2, Escola de Cultura de Pau, 2008); Hilary Charlesworth and Mary Wood, "Women and Human Rights in the Rebuilding of East Timor," *Nordic Journal of International Law* 71 (2002): 325–48; Kvinna till Kvinna, *Getting it Right? A Gender Approach to UNMIK Administration in Kosovo* (Stockholm, Sweden: Kvinna till Kvinna, 2001); UN Department of Peacekeeping Operations, *Gender Mainstreaming in Peacekeeping Operations: Progress Report* (New York: United Nations, 2005).

21 UN Department of Peacekeeping Operations, *Gender Mainstreaming in Peacekeeping Operations: Progress Report.*

22 Lori Handrahan, "Conflict, Gender, Ethnicity and Post-Conflict Reconstruction," *Security Dialogue* 35, no. 4 (2004): 429–45; Donna Pankhurst, "Introduction: Gendered War and Peace," in *Gendered Peace: Women's Struggles for Post-War Justice and Reconciliation*, ed. Donna Pankhurst (New York: Routledge, 2008), 1–30.

23 Garth Massey, Karen Hahn, and Duško Sekulić, "Women, Men, and the 'Second Shift' in Socialist Yugoslavia," *Gender & Society* 9, no. 3 (1995): 359–79; Sabrina P. Ramet, "In Tito's Time," in *Gender Politics in the Western Balkans: Women and Society in Yugoslavia and the Yugoslav Successor States*, ed. Sabrina P. Ramet (University Park, PA: The Pennsylvania State University Press, 1999), 89–105; Tanja Rener, "Yugoslav Women in Politics: Selected Issues," *International Political Science Review* 6, no. 3 (1985): 347–54.

24 Ramet, "In Tito's Time," 95.

25 Massey, Hahn and Sekulić, "Women, Men, and the 'Second Shift' in Socialist Yugoslavia," 359–79; Mary K. Gilliland Olsen, "Redefining Gender in Yugoslavia: Masculine and Feminine Ideals in Ritual Context," *Eastern European Quarterly* 23, no. 4 (1989): 431–44; Susan L. Woodward, "The Rights of Women: Ideology, Policy, and Social Change in Yugoslavia," in *Women, State, and Party in Eastern Europe*, ed. Sharon L. Wolchik and Alfred G. Meyer (Durham, NC: Duke University Press, 1985), 234–56.

26 Barbara W. Jancar, "Neofeminism in Yugoslavia: A Closer Look," *Women & Politics* 8, no. 1 (1988): 1–30.

27 Global Rights (in cooperation with NGOs in BiH), *Shadow Report on the Implementation of CEDAW and Women's Human Rights in Bosnia and Herzegovina* (January 2004), http://www.rightsforall.ba/eng/dw/CEDAW_Shadow_Report_BiH_Final_2004.pdf; Alexandra Petrić, Fedra Idžaković, Gordana Vidović, Natalja Petrić, Mara Radovanović, and Diana Šehić, *Alternative Report on the Implementation of CEDAW and Women's Human Rights in Bosnia and Herzegovina* (October 2010), http://www.zenezenama.org/eng/CEDAW1.pdf; UNDP BiH, *National Human Development Report: Social Inclusion in Bosnia and Herzegovina* (2007), http://www.undp.ba; *Combined Initial, Second and Third Periodic Reports of States Parties: Bosnia and Herzegovina* (Committee on the Elimination of Discrimination against Women document CEDAW/C/BIH/1–3), 18 April 2005.

28 For a comprehensive list see Jenichen, *Mehrebenen-Advocacy-Koalitionen in internationalisierten Nachkriegsstaaten*, chapter 2.1.

29 Cynthia Cockburn, *The Space Between Us: Negotiating Gender and National Identities in Conflict* (London: Zed Books, 1998); Djurić-Kuzmanović, Drezgić, and Žarkov, "Gendered War, Gendered Peace," 265–91; Maja Korac, "Gender, Conflict and Peace-building: Lessons from the Conflict in the Former Yugoslavia," *Women's Studies International Forum* 29 (2006): 510–20.

30 Kvinna till Kvinna, *To Make Room for Changes: Peace Strategies from Women's Organisations in Bosnia and Herzegovina* (Johanneshov, Sweden: Kvinna till Kvinna, 2006).

31 GEEP, *Gender Equity and Equality Project Completion Report* (Sarajevo, BiH: unpubl., 2006); Anne Jenichen, "Frauenorganisationen und externe Friedensförderung in Bosnien-Herzegowina," *Femina Politica* 16, no. 2 (2007): 72–82. The willingness of external donors to promote norm and policy entrepreneurs aiming at challenging predominant political discourses and mobilizing political support for their concerns has been generally low in BiH. See Adam Fagan, "Civil Society in Bosnia Ten Years After Dayton," *International Peacekeeping* 12, no. 3 (2005): 406–19.

32 Open Society Institute, *On the Road to the EU: Monitoring Equal Opportunities for Women and Men in South Eastern Europe: Overview* (Budapest, Hungary: Network Women's Program, 2006), http://www.soros.org/initiatives/wo

men/articles_publications/publications/onroad_20060503/aa_onroad_200605 17.pdf, 48.

33 Kvinna till Kvinna, *Engendering the Peace Process.*

34 Besima Borić, "Application of Quotas: Legal Reforms and Implementation in Bosnia and Herzegovina," paper presented at the International Institute for Democracy and Electoral Assistance (IDEA)/CEE Network for Gender Issues Conference, The Implementation of Quotas: European Perspectives, Budapest, 22–24 October 2004.

35 The OHR is the international transitional administration in BiH, overseeing the implementation of the Dayton Peace Agreement.

36 Global Rights, *Legislative Advocacy Resource Guide: Promoting Human Rights in Bosnia and Herzegovina* (Sarajevo, BiH: Global Rights, 2005).

37 It must be noted here that this organization, although registered as an international NGO, had exclusively been run by domestic staff since 1997. Many domestic women's rights activists, therefore, perceived it rather as a domestic NGO. However, the resources it had at its disposal differed from the resources domestic NGOs possess.

38 Kathryn Sikkink, "Patterns of Dynamic Multilevel Governance and the Insider-Outsider Coalition," in *Transnational Protest and Global Activism*, ed. Donatella della Porta and Sidney G. Tarrow (Lanham, MD: Rowman & Littlefield, 2005), 151–73.

39 Susanne Alldén, "Post-Conflict Reconstruction and Women's Security: Analyzing Political Outputs in East Timor," *Umeå Working Papers in Peace and Conflict Studies*, no. 1 (2007); Susanne Alldén, "The Aftermath of International Administrations: Socializing Women's Political Rights in Cambodia," paper presented at the International Studies Association Annual Convention 2009, New York, 15–18 February; Carol Cohn, "Mainstreaming Gender in UN Security Policy: A Path to Political Transformation?" in *Global Governance: Feminist Perspectives*, ed. Shirin Rai and M. Georgina Waylen (New York: Palgrave Macmillan, 2008), 185–206; Jasmina Džumhur, "Election Law of B&H in the Light of Gender Equality Law in B&H," *Woman Today*, no. 9 (2006): 136–44; Global Rights, *Shadow Report on the Implementation of CEDAW and Women's Human Rights in Bosnia and Herzegovina*; Ane Jakšić, "For Equal Possibilities of Both Sexes in the Election Law in B&H," *Woman Today*, no. 9 (2006): 145–54; Kosova Women's Network, *Monitoring Implementation of UNSCR1325 in Kosovo: Executive Summary* (2007), http://www.eulex-kosovo.eu/training/material/docs/KR/KW N_Exective_Summary_on_the_Implementation_of_UNSCR1325.pdf; Žene ženama, *Final Report: Participation of the Public in Human Security: UN SCR 1325 in Bosnia and Herzegovina* (2010), http://www.eupm.org/FCKedi tor/Images/File/Final_Report_1325_UNSCR.pdf.

40 Gallup Balkan Monitor, *Focus on: Perceptions of the EU in the Western Balkans* (2009), http://www.balkan-monitor.eu/files/Gallup_Balkan_Monito r-Focus_On_EU_Perceptions.pdf; International Crisis Group, "Ensuring Bosnia's Future: A New International Engagement Strategy," *Europe Report*, no. 180 (2007), http://www.crisisgroup.org/~/media/Files/europe/ 180_ensuring_bosnias_future.pdf.

41 Ana E. Juncos, "The EU's post-Conflict Intervention in Bosnia and Herzegovina: (Re)Integrating the Balkans and/or (Re)Inventing the EU?" *Southeast European Politics* 6, no. 2 (2005): 88–108.

42 Sevima Sali-Terzić, "Civil Society," in *International Support Policies to South-East European Countries: Lessons (Not) Learned in Bosnia-Herzegovina*, ed. Žarko Papić (Sarajevo, BiH: Open Society Fund BiH, 2001), 175–94.

43 Jenichen, *Mehrebenen-Advocacy-Koalitionen in internationalisierten Nachkriegsstaaten*, chapter 7.

Part IV

Economic governance

12 Governing the economy for gender equality? Challenges of regulation

Shahra Razavi

- **Good governance: its promises and limits**
- **The return to "the social": mothers, children and the family**
- **Diffusing ideas to the Global South**
- **Governing the market for gender equality?**
- **Conclusions**

Four years into the 2007/8 global economic crisis and with the semblance of a return to "business-as-usual" that leaves the deeper structural causes of the crisis unattended, the question of economic governance looms large.[1] The crisis may have opened up a small democratic space for short-term protective measures and some talk of longer-term regulatory interventions to prevent similar collapse and bailout in the future. However, "this space was equally quickly shut down again, as the political and financial establishment resisted yielding permanent ground."[2] As Bob Jessop argues, there are always many different interpretations of the crisis and of how to respond to it, but there are also asymmetries of power which are significant in the selection of crisis-interpretations and their translation into crisis-responses. This is a key factor behind the re-assertion of key elements in the hegemonic (neoliberal) project despite the initial shock to that project from the current crisis.[3] The response in the rich industrialized countries—the return to fiscal conservatism (especially in Europe) and to the bonus bonanzas of late—reflect the anti-democratic nature of economic governance in some of the world's oldest and most consolidated democracies. Some see this as evidence of a slippage into plutocracy.[4] Rising levels of income inequality contribute to the erosion of institutional norms and behavior and the subversion of government, as those with economic power use "all the instruments that extreme wealth puts at their command to protect and perpetuate their position in economy, polity and society."[5]

Would more democratic forms of economic governance have produced different outcomes, more economic security and social equality?

Would the structural malaise that underpinned the crisis—the rising income inequalities linked to stagnant wages and rising profits, and rampant financialization of the economy—have been addressed sooner and more effectively? Would the risks and costs of crisis-management have been distributed more fairly, rather than being pushed onto subaltern groups, among them many women?[6] These are difficult questions to answer but ones that have already prompted thinking and debate within both activist and academic circles. This paper, along with the others in this part, examines a more specific aspect of that debate, namely *what kind of economic governance do we as feminists want?*

While I take the 2007/8 crisis as my point of departure in terms of the key questions that it raises, the chapter adopts a longer time perspective on the question of economic governance, looking back at the dynamic interaction between feminist claims about social costs of economic restructuring and policy responses to those claims.

The rest of the chapter is structured as follows. In the section following this introduction the notion of economic governance is explained and its current policy usage ("good governance") problematized for its gender-specific blind spots, rooted in the tendency to abstract "the economic" from "the social," and to neglect equality and justice issues as values and goals. Section two then begins by recalling the critique of the orthodox agendas of the 1980s, which feminists argued were deeply gendered both in their assumptions and outcomes. It examines some of the ideational changes that subsequently took place, initially at the European level as welfare advocates sought to re-legitimize social policy and make it more palatable within a largely "market-friendly" policy environment. Interestingly, children, women and the family have figured strongly in these post-neoliberal policy propositions. The diffusion of some of these ideas into globalized policy circuits and their filtering into the very different socio-political contexts of developing countries, it is argued, has also brought out some of their limitations, not only for developing country contexts but also the "developed" countries that were their initial points of reference. The final section returns to the intellectual silos that continue to attribute gender inequalities to the workings of "private sphere" institutions, and asks whether we are likely to move beyond this impasse.

Good governance: its promises and limits

In the 1980s the World Bank did not have much to say directly about governance, the assumption being that once market prices were set right, the state would be efficient in its task of rule enforcement, protection of property rights and public administration. However, the disappointing economic performance of countries undergoing World Bank-tailored reforms, especially in sub-Saharan Africa, led to a new interest in, and indeed discovery of, the role of institutions in promoting development. This seemed like a welcome move, for it had the potential to draw attention to the relationship between patterns of development management and development outcomes. However, the new interest in governance and in institutions, rather than prompting new departures and diversifying strategies and pathways, turned out to do much less.

First, the "good governance" agenda, as it was subsequently called, aimed to facilitate the interaction between economic agents and the state in a way that was market-enhancing. Hence, what made it "good" was the management of the economy to produce market efficiency and growth; very little was said about equity and social justice.[7] Second, instead of expanding the range of institutional strategies to be explored, the most prominent policy outcome of this "institutional turn" was the rise of what Peter Evans calls "institutional monocropping": the imposition of blueprints based on idealized versions of Anglo-American institutions, the applicability of which is presumed to transcend national circumstances and cultures."[8]

Good governance reforms have thus emphasized the crafting of institutions to promote the "rule of law," protect property rights, reduce levels of corruption and improve regulatory quality. In sub-Saharan Africa, for example, the overwhelming emphasis and the bulk of investment under the rule of law rubric has gone into creating a suitable legal and institutional environment for the market, through reforms in areas such as commercial codes, bankruptcy, banking, tax and property laws including intellectual property, freedom of information and so forth.[9] Other items on the agenda have included the creation of independent central banks and autonomous tax boards to ensure "sound" macroeconomic management.

The expectation has been that by promoting this type of institutional change a country's economic performance would be enhanced. Such expectations, however, have not been met. One central question that development economists have asked is whether expanding the efficiency of markets will be sufficient to maximize the pace of development. Empirical tests carried out by Mushtaq Khan on each of the good governance indicators—property rights, regulatory quality, corruption, voice and accountability—show the role of market-enhancing governance conditions in explaining growth rates in developing countries to be weak.[10] The key implication of these findings is that other types of governance capabilities, ignored by the good governance agenda, may in fact be the ones that make a significant difference in developing countries' ability to register good rates of growth and overcome economic backwardness.

Social activists and other critics stress that the effectiveness of any state will be determined in great part by its legitimacy and thus stress that governance is only as "good" as the manner in which power is acquired and eventually surrendered by rulers, and used to promote social justice. Gender equality, like broader issues of equality, has not been a fundamental concern of good governance reforms. In the World Bank's early statements concerning governance reform, women's participation was explicitly addressed only in discussions of decentralization and the use of participatory approaches; and gender equality was only raised in the context of human rights. Yet there are gender-specific capacity failures in all public institutions targeted for reform—for example ending corruption and institutional bias (including gender bias) in the functioning of institutions that administer justice, or correcting public expenditure management systems that fail to acknowledge women's needs and distribute resources equitably. In practice the gains for gender equality through such reforms have been few and far between. Other features of governance reform that involve privatization of essential public services, decentralization of decision-making and informalization of dispute resolution mechanisms are producing new forms (or re-authorizing "traditional" forms) of patriarchal social control.[11]

This is not, however, the end of the story. Besides governance reforms, the "post-Washington consensus" has also embraced poverty reduction and social protection programs. These proposals ostensibly seek to mend the ravages of harsh stabilization and neoliberal restructuring, which feminists claimed had shifted the costs of social reproduction to families and households, with a particularly detrimental impact on women. The turn to poverty and social protection thus seemed to suggest that "development agencies and governments were finally taking on board more than thirty years of criticism of their top down, 'growth first' policies."[12] There is growing adherence to the view that if economic liberalization is to stay on course and not be derailed by social protest and simmering unrest then it needs to be "embedded" through social policies and anti-poverty programs.[13] The latter have therefore mushroomed across a wide range of countries. Within these programs women, children and the family have emerged as visible targets and recipients.

Hence, the social deficit and discontent emerging from patterns of economic development which are largely exclusionary and polarizing are now being addressed in the messy realm of social policy formulation and implementation. Here political elites (sometimes in conjunction with external actors) interpret, appease, deflect or subvert the articulated "needs" of subaltern groups, including women's rights advocates, through a mix of regulatory and provisioning measures. How well are gender claims for equality being taken up? The results have been mixed. Despite some rhetorical claims, the attention to *gender equality* has been uneven. Women's claims have sometimes been subsumed under children's needs and capabilities—a major preoccupation of the social investment perspective which has risen to global dominance over the past decade.[14] Efforts to address women's needs and interests in the here-and-now seem to have fallen by the wayside.

The return to "the social": mothers, children and the family

It does not take fancy econometric models to show that the time and effort that people put into taking care of others—whether this is mediated through kinship and family ties, neighborly relations and friendship, or through paid forms of care—constitutes a key input into the well-being of not only children and frail elderly persons, but also able-bodied adults. But it took a lot of work by feminist writers and activists to expose the "hidden, taken-for-granted unpaid caring and domestic work that women did in the home."[15] Their voices were joined by the care claims of other social movements, including disability rights activists who contested the social exclusion of disabled and old people, carers' claims for recognition of their work, and children's rights activists articulating children's rights to good quality care.

In Europe these calls began to be taken more seriously as patterns of social change brought more women into the paid workforce, as family and household arrangements diverged from the normative model (of heterosexual married couple with children), and in some contexts, as concerns grew about falling fertility rates and the difficulties this would pose for keeping pension systems solvent. For some analysts these changing demographic and social patterns required a recalibration of modern welfare states away from their historical preoccupation with the risks of old age (pensions) and unemployment visited upon the male breadwinner and his family, to the "new social risks" associated with post-industrial societies and economies which had undergone significant change. Gender has figured as a major theme in this literature, as the title of Gøsta Esping-Andersen's book attests: *The Incomplete Revolution: Adapting Welfare States to Women's New Roles.*[16] The "new" social risks have been variously defined to include the reconciliation of work and family life; single-parenthood; caring for frail relatives; possessing low or obsolete skills; and insufficient social security coverage.[17] Assisting families and women to reconcile their work and family responsibilities is seen as the surest way of combating poverty and its intergenerational transmission while also allowing couples to have the number of children they desire (where the Swedish model is said to have worked, while the southern European and East Asian ones have allegedly failed).

These policy responses, however, were far from automatic. To shift policy away from the retrenchment/austerity mindset of the 1980s, to one that was responsive to changing social and economic realities, perhaps even transformative, required new ideas that were simply not on offer on the neoliberal menu. But by the mid-1990s spaces were opening up for new thinking. Jane Jenson has traced how policy-makers at the international, regional and national levels began to converge around the key principles of what is now widely referred to as the "social investment perspective."[18]

Central to this post-neoliberal perspective is the notion of the state as an investor: to be effective state spending must not be consumed in the present, to ameliorate current conditions; it must be an investment that will reap rewards in the future. "In other words this involves going from a welfare state that is a 'nurse' to one that is an investor."[19]

Closely linked to the idea of the state as an "investor," the new perspective has also implied a different notion of time to that found in the post-war welfare state. The post-1945 welfare state was strongly focused on the task of redistribution, on fostering greater equality in the *here-and-now*. The social investment state, by contrast, puts the accent on the "equality of life chances" which, as Jenson explains, involves distribution and redistribution of opportunities and capabilities, but not necessarily of resources. As such the social investment perspective is fully in tune with the notion of "equality of opportunity" that has become part of the common-sense in current thinking about equality, displacing the earlier concern with equality of outcome.[20]

A third element is the notion of learning as the pillar of the economies and societies of the future (knowledge-based economies). Coming together these ideas have lent themselves to a particular focus on children, their human capital and capabilities, and contributed to a concerted push for child-centered interventions.[21] More problematically though, some feminists argue that this child-centered policy focus has effectively sidelined the equality claims of adult women; their needs and interests have been "lost in translation," sidelined in favor of those of children, and in particular those of girls.[22]

In Europe, there are a number of major thrusts to reform, and countries have been steered by the European Union (EU) and the Organisation for Economic Co-operation and Development (OECD) towards a policy on childcare. The EU has been particularly influential in shaping policies for work/family reconciliation through parental leaves and childcare services. Unlike the World Bank, which has advocated targeted programs for the poor, the EU and OECD have placed the accent on universal and institutionalized forms of early childhood education and care services (ECEC). However, the EU's interventions have been predominantly framed by a concern to encourage women's labor market "activation" within a "flexible" employment regime that is necessary for increasing competitiveness and growth. Maria Stratigaki's analysis of relevant EU acts shows how a concept introduced to encourage gender equality in the labor market, the "reconciliation of working and family life," gradually shifted in meaning from an objective with feminist potential ("sharing family responsibilities between women and men") to a market-oriented objective ("encouraging flexible forms of employment") as it became incorporated in the European Employment Strategy of the 1990s. She characterized this process as one of co-optation because the goals of the original proposals are undermined by shifting the meanings of the original concepts to fit into the prevailing political and economic priorities.[23] Others contend that the economic imperative to increase female employment has resulted in some "extraordinarily instrumentalist arguments" which do not problematize the kinds of jobs that women are getting; nor do they deal with the question of gender inequality in unpaid work, given the limits of de-familialization as a strategy (after all, it is very difficult to de-familialize *all* unpaid care work); finally, the policy positions adopted do not provide any space for women to make real choices about care.[24]

For the OECD, too, the main frame has been female labor market participation to combat lone parent poverty and to arrest falling birth rates (as reflected in the prominent series, *Babies and Bosses*, issued by the OECD Directorate on Employment).[25] In short, slowly but surely gender equality has become "a minor frame in relation to the dominant frame of employability."[26]

Diffusing ideas to the Global South

Within development circles until recently a dominant view was that allocating public resources to social policy is "unproductive" and that it "crowds out" private savings and investment, and can thus become an impediment to growth. These ideas shaped the structural adjustment programs of the 1980s that cut deeply into public social expenditure in many developing countries. However, since the mid-1990s a more enabling set of ideas have started to question the dominant viewpoint. Drawing on a wide range of ideas, including the social investment perspective, attention has been drawn to the "productive" role of social policy in economic development.[27] Not only is social policy good for redistribution and protection, but it can also make vital contributions to innovation, which is critical for developing countries' efforts to "catch-up" with the developed world.[28]

Jenson shows how many of the ideas associated with the social investment perspective have spread in Latin America, even though not all jurisdictions have embraced the ideas "at exactly the same time or with the same enthusiasm."[29] Nevertheless the notion of social spending as "investment," the emphasis on children's capabilities and investments therein to break the intergenerational cycle of poverty, and the assignment of responsibility to the state (within the welfare mix) for ensuring that such investments are possible, are some of the common features that she identifies.

Indeed, children and child poverty have been prominent themes in development debates and policies of the past couple of decades. The role played by the United Nations Children's Fund (UNICEF) in issuing the first "wake-up call" from within the UN system in response to the austerity measures imposed by the international financial institutions (*Structural Adjustment with a Human Face*), and the widespread ratification of the 1989 Convention on the Rights of the Child were important moments in defining this new focus. Within the framework of the Millennium Development Goals (MDGs) that is supposed to steer both donor and government policy efforts, tellingly, gender equality (Goal 3) is being operationalized through a target that is about closing the gender gap in primary and secondary education—the focus being on girls (or the "girl child"), while the only goal that concerns adult women is about improving maternal health (Goal 5).

Yet if the social investment approach has in one context, Europe, sought to push women into the labor market under the theme of "employability," elsewhere, especially in Latin America, critics contend that it has naturalized motherhood and reinforced the maternalist thrust in social policy.[30]

A new generation of social assistance programs, known as conditional cash transfer (CCTs) schemes, are specifically targeted to poor women in their capacity as mothers and household managers, and made conditional on compliance with certain requirements linked to children's well-being (such as regular health checks, and the participation of mothers in nutrition workshops). Feminist critics contend that while these programs have some potential to support women in their role as caregivers, they do not broaden their options beyond that role, by giving them a more secure footing in the labor market for example. Some have even argued that CCTs reinforce the traditional gender division of labor by discouraging men from care-giving, while they overburden women whose (paid and unpaid) inputs into household survival have diversified and intensified in many developing countries.

CCTs, however, represent one thrust of recent policy efforts aimed at children, but not the only one. Middle-income developing countries such as Argentina, Chile, Mexico, South Africa and Uruguay have been experimenting with a range of care-related social policies, including ECEC which can potentially be more supportive of women's movements demands by expanding women's options. All of these countries are also characterized by high levels of income inequality. The challenge they face therefore is not only to expand service coverage, but to do so in a way that reduces class and regional inequalities in the quality of service accessed by children from different socio-economic groups. This becomes a formidable challenge when a mix of public and private provision is used, and where different kinds of "public" services are targeted to children from different backgrounds.

To give some concrete examples, class and regional differences in access to preschool education for five-year-olds have been substantially reduced in Argentina by making enrollment mandatory for this age group and by expanding public preschools. However, for the younger age groups where public provision is limited and the market plays a dominant role, enrollment rates for children from lower-income households remain only a fraction of those of children from higher-income groups. Since low-income families cannot afford private childcare, they face long waiting lists for public crèches, often rely on less professionalized "community" services or on unpaid care by family members.[31]

A similar situation prevails in Mexico, where from the early 2000s the Ministry of Public Education made public preschool education mandatory for all three- to five-year olds, while the Ministry of Social Development (Sedesol) has, since 2007, put in place a large daycare program (Federal Daycare Program for Working Mothers) that is targeted to a younger cohort of children (one to four years) from low-income households; the Federal Daycare Program also relies on less professionalized and lower-cost care services.[32] Interestingly and in some contrast to the maternalist thrust of the Mexican cash transfer program to which reference has already been made, the Mexican Federal Daycare Program's main aim has been to expand employment opportunities for women (rather than enhancing child welfare or school-readiness). It claims to do this by giving mothers access to affordable community-based childcare options, and by creating employment opportunities for women as self-employed carers offering their services under the program.

Whether aimed at mothers' labor market activation (the Mexican program) or building the capabilities of children and making them school-ready (the Argentine program), the recent expansion of ECEC services also raises a number of concerns.

The first set of concerns relates to the *segmented nature of such programs*. The fact that services are provided by a mix of public, private and "community" providers (the latter very often meaning less professionalized and cheaper services, with lower staff/child ratios and fewer facilities and materials), targeted to different social groups, complicates the task of equalizing the quality of services/benefits while creating the danger of parallel and unequal provision for different groups. It is difficult to talk about "equality of opportunity" when both children and their family carers face such different care options. Higher-income groups are provided with good quality care services (the gaps between opening hours and parents' working time filled by domestic workers) that facilitate their growth and flourishing, and their carers have the option of holding on to stable employment. Among the poor, care has to be abbreviated as family carers seek irregular and unprotected paid work, and the services they access are often of questionable quality.

The second issue is about the kind of jobs that are being created, especially for women in these and many other care-related social programs. De-familialization of care (i.e., shifting care out of the family) is now advocated by a wide range of policy advocates as a mechanism for expanding employment opportunities for women—an important pillar, we are told, of the Swedish model.

The first obvious point to make is that de-familialization of care and its commodification have not changed the fact that it is still predominantly women who do this kind of work (even in Sweden!). Moreover, we know from a wide range of research on both developed and developing countries that care work often carries a pay penalty in terms of wages and salaries regardless of the level of skill and education content of the work.[33] What is particularly problematic is that the recent expansion of care-related social programs relies heavily on "voluntary" or "community" work. This may be cost-effective for the state. But that is because the costs are shifted to the largely feminized workforce that delivers these services, many of them unpaid or underpaid "volunteers." In India, for example, the *Anganwadi* workers and helpers who staff the Integrated Child Development Scheme—probably the largest early childhood nutrition scheme globally—fall into this ambivalent category. While employed by the state, they are *not* classified as workers, but as "volunteers" who receive stipends instead of wages and lack the leave entitlements and social security benefits available to permanent, full-time public employees.[34]

This is not a trend that is limited to the developing world. As Marilyn Waring puts it in an interview commenting on the devolution of government services in New Zealand to "the community": "We know that 'community' is usually mom or daughter or aunty or neighbor or some other woman who already works 16 to 18 hours a day. Sometimes she's in the paid work force; sometimes not."[35] A decade later, Waring's statement resonates in the post-2008 context of fiscal retrenchment, where the notion of "Big Society" is invoked to fill the gap left in the wake of "state rollback" in countries such as the United Kingdom.

What happened to women's equality claims, their concerns with wage discrimination and gendered hierarchies of work? These gender inequalities and discriminatory practices are being constantly reproduced through the broader workings of the liberalized capitalist economy. The emphasis on "human capital," education, and the preparation of future workers for successful labor market participation—the key concerns of the social investment state—are after all based on a "supply-side" analysis. Yet, as we know from decades of feminist research, the inequalities that women face are not because of low levels of education or preparedness for work. They are in fact due to systemic and structural ways in which markets and the economy work.

Governing the market for gender equality?

The effort to re-embed the economy through social policy has taken place without abandoning the neoliberal basics centered on economic liberalization, fiscal restraint, and a nimble state that facilitates the integration of people into the market. This has very often meant the continued reproduction of gender and other social inequalities through markets and macroeconomic policies.

Unequal power relations of gender pervade both the public (markets and states) and the private institutions (of family, households and community). Yet the World Bank continues to portray markets and macroeconomic flows as essentially benign and gender-neutral. Gender equality/inequality/difference is only recognized as and when it operates within the institutions of the domestic sphere, and what the World Bank calls "customary" and "traditional" practices.[36] Land markets, for example, have been enthusiastically proposed as a solution for women smallholders in Africa and elsewhere, side-stepping evidence that shows land markets to be severely limited as a channel for women's inclusion.

At the same time very little is said about the very low wages of women agricultural laborers, and the discrimination and harassment to which they are subject. The long tradition of feminist work that has analyzed labor markets as *social* and gendered institutions is thus effectively brushed aside. In short, there is little recognition of the ways in which "gender inequality is not only weakened but also recreated and sustained by capitalist development … that a considerable degree of inequality is 'hard-wired' into the functioning of capitalist economies."[37]

In the aftermath of the 2007/8 crisis much has been said about the need for stronger regulation of financial systems (see also contribution by Young in this volume). But such prudential measures need to be supplemented by other policies and measures: having employment objectives at the heart of macro-economic policy; and nurturing, through industrial policies and the like, what economists refer to as "wage-led" economies (i.e., economies where production is oriented toward supplying a broad-based domestic market based on rising labor productivity and rising wages) rather than export-oriented economies where wage increases lag behind productivity increases. It also requires the active management of cross-border flows of trade and capital, rather than simply pursuing trade and financial liberalization as an end in itself.

There also needs to be much better recognition of how liberalized markets reproduce inequalities. What are we to make of the fact that a big and growing part of the labor force operates beyond the regulatory reach of the state, in what is frequently referred to as the informal economy? As Barbara Harriss-White rightly observes, the absence of state regulation in this part of the economy does not mean that there is no regulation. No market exists in a regulatory vacuum.[38] Effectively, then, "deregulation" has meant greater freedom for business and markets *from* organized labor. At the same time the social institutions of gender, caste and other forms of identity operate as critical *social regulators* in this growing part of the economy. Social regulation creates a regulative order in markets as well as shaping many practices of well-being and security. But it is a highly hierarchical and discriminatory order, policed by local elites and "strongmen" (often in collusion with party bosses and local-level state bureaucrats). For women it means discriminatory wages, provision of minimal standards of decency and safety at work, and no access to social security in many contexts.

Upsetting this "order" is therefore a hugely political project. But it is one that has been postponed for too long—the neoliberal argument was that growth in itself would "lift all boats." In practice labor informality grew and the relative share of wages fell, even in contexts where growth rates were high, as in India and China. It does not require much research to see the desirability of having a set of clear "public policy choices which create, expand and give teeth to, the legal rights available to workers regardless of gender, caste etc. both at work and with respect to social security and protection."[39] We know from the failure of the legal-determinist project of law and development in the 1970s that the politics of labor rights is "riddled with strategic slippages"—between commitments made to UN conventions and those codified in national laws and procedures; between national laws and "on-the-ground" implementation. What is clearly needed are the social forces to back the instituting of rights that have been on statute books. Policy advocates of women's rights, including UN Women, need to push forward efforts to engender economic governance both conceptually and in practice while supporting the social forces striving to make economic rights a reality.

Conclusions

This chapter has shown how economic governance has tended to evade claims for gender equality, or has responded to them only very partially. Using the most recent episode of crisis in the hegemonic neoliberal model as its point of departure, it looked back to the dynamic interaction between the feminist critique of neoliberal economic restructuring and policy responses to those claims during the "post-Washington Consensus" era. It showed the emergence of new ideas about social policy and the role of the state ("the social investment state") in this period, initially within the European context, and their diffusion across space to global and developmental policy circles. The strong interest in the development of children's capabilities on the part of the "social investment state" has resonated in particular with the longer-standing development concerns with children's rights and welfare. Child-centered "social investment" policies have taken various forms and involved women in a variety of capacities—as recipients of cash transfers in their capacity as mothers/carers, but also as workers delivering services in ECEC programs. Despite some of the tangible benefits that these programs can deliver to their beneficiaries, they are also directly implicated in the reproduction of social inequalities. Patterns of class/regional disadvantage as far as children are concerned are often reproduced through segmented social programs (delivering different quality of service to different social groups). The reliance on the unpaid or underpaid work of women also serves to reproduce gender-based inequalities within the labor market. The latter, we have argued, is in fact part of the much broader pattern of economic governance that has allowed a growing part of employment to operate beyond the regulatory reach of the state. To address informality and gender-based hierarchies within labor markets, and to facilitate the translation of equality legislation into every-day practices, requires organized constituencies operating at multiple levels (local, national, regional, global).

Notes

1 By economic governance we mean the institutional arrangements—the rules and norms, both formal and informal—that determine the respective roles and responsibilities of states, markets, and families and shape how economies perform (in terms of growth, distribution and meeting welfare needs).

2 Ashwani Saith, "Inequality, Imbalance, Instability: Reflections on a Structural Crisis," Forum 2010/2011, *Development and Change* 42, no.1 (2011): 70–86, 70.

3 Bob Jessop, "Narratives of Crisis and Crisis Response: Perspectives from North and South," in *The Global Crisis and Transformative Social Change*, ed. Peter Utting, Shahra Razavi and Rebecca Buchholz (Basingstoke: UNRISD/Palgrave, forthcoming).

4 Robert Reich, "Reading America's Tea Leaves," *The American Interest* VI, no.2 (2010): 6–17.

5 Saith, "Inequality, Imbalance, Instability," 71.

6 The gender implications of the budget cuts in the United Kingdom have been analyzed by the UK Women's Budget Group. Their rigorous analysis is sobering: it demonstrates that fiscal retrenchment not only has a strong class dimension but also an unmistakable gender content. UK Women's Budget Group, *A Gender Impact Assessment of the Coalition Government Budget* (London: Women's Budget Group, 2010).

7 Anne-Marie Goetz, "Governing Women or Enabling Women to Govern? Gender and the Good Governance Agenda," in *Governing Women: Women's Political Effectiveness in Contexts of Democratization and Governance Reform*, ed. Anne-Marie Goetz (New York: Routledge/UNRISD, 2009), 239–56.

8 Peter Evans, "Development as Institutional Change: The Pitfalls of Monocropping and the Potentials of Deliberation," *Studies in Comparative International Development* 38, no. 4 (2004), 30.

9 Celestine Nyamu-Musembi, *For or Against Gender Equality? Evaluating the Post-Cold War "Rule of Law" Reforms in Sub-Saharan Africa*. OP7 (Geneva: UNRISD, 2005).

10 Mushtaq H. Khan, "Governance, Growth and Development," background paper for UN Research Institute for Social Development (UNRISD) report on *Combating Poverty and Inequality* (Geneva, UNRISD, 2008) cited in UNRISD, *Combating Poverty and Inequality: Structural Change, Social Policy and Politics* (Geneva: UNRISD, 2010). The UNRISD report provides its own analysis of what these other governance capabilities are: the capacity to forge political settlements for redistribution; improving domestic resource mobilization and encouraging citizen involvement in the allocation and monitoring of resources; improving state capacity for service delivery; and improving the quality of decentralization by involving

subaltern groups in local decision-making processes. Unfortunately, the gender specificities of these capacities are not elaborated.

11 Goetz, "Governing Women or Enabling Women to Govern?" 239–56.

12 Maxine Molyneux, "Gender and the Silences of Social Capital: Lessons from Latin America," *Development and Change* 33, no. 2 (2002): 167–88, 172.

13 The concept of "embedded liberalism" was initially coined by John Ruggie to refer to the institutional arrangements put in place after World War II in many Western capitalist countries, to share the costs of open markets, through social protection and other forms of intervention (such as wage restraint by trade unions in return for employment creation on the part of capital). The notion drew on Karl Polanyi's ground-breaking analysis of the "double movement" in the early years of industrialization in Britain when the social disruptions brought about by unregulated market forces, or the "disembedding" of markets from society, gave rise to the need for a "counter-movement" to bring markets under societal control. John G. Ruggie, "Taking Embedded Liberalism Global: The Corporate Connection," in *Taming Globalization: Frontiers of Governance*, ed. David Held and Mathias Koenig-Archibugi (Cambridge: Polity Press, 2003), 93–129; Karl Polanyi, *The Great Transformation* (Boston, MA: Beacon Press, 1957).

14 Jane Jenson, "Diffusing Ideas for After Neoliberalism: The Social Investment Perspective in Europe and Latin America," *Global Social Policy* 10, no. 1 (2010): 59–84.

15 Fiona Williams, *Claims and Frames in the Making of Care Policies. GD Programme Paper No.13* (Geneva: UNRISD, 2010): 2.

16 Gøsta Esping-Andersen, *The Incomplete Revolution: Adapting Welfare States to Women's New Roles* (Cambridge: Polity Press, 2009).

17 Guiliano Bonoli, "New Social Risks and the Politics of Post-Industrial Social Policies," in *The Politics of Post-Industrial Welfare States: Adapting Post-War Social Policies to New Social Risks*, ed. Klaus Armingeon and Giuliano Bonoli (New York: Routledge, 2006).

18 Jane Jenson and Denis Saint-Martin, "Building Blocks for a New Social Architecture: The LEGO Paradigm of an Active Society," *Policy and Politics* 34, no. 3 (2006): 429–51; Jane Jenson and Denis Saint-Martin, "New Routes to Social Cohesion? Citizenship and the Social Investment State," *Canadian Journal of Sociology* 28, no. 1 (2003): 77–99.

19 Bruno Palier cited in Jane Jenson, "Lost in Translation: The Social Investment Perspective and Gender Equality," *Social Politics* 16, no. 4 (2009): 447.

20 Anne Phillips, "Really Equal: Opportunities and Autonomy," *The Journal of Political Philosophy* 14, no. 1 (2006): 18–32.

21 Jenson and Saint-Martin, "Building Blocks for a New Social Architecture," 429–51.

22 Jenson, "Lost in Translation," 446.

23 Maria Stratigaki, "The Cooptation of Gender Concepts in EU Policies: The Case of Reconciliation of Work and Family," *Social Politics* 11, no. 1 (2004): 30–56.

24 Susy Giullari and Jane Lewis, *The Adult Worker Model Family, Gender Equality and Care: The Search for New Policy Principles and the Possibilities and Problems of a Capabilities Approach. SPD Programme Paper No. 19* (Geneva: UNRISD, 2005): 6.

25 Mahon shows, however, that other OECD divisions have paid more attention to issues of gender equality for adult women. Rianne Mahon, "After Neoliberalism? The OECD, the World Bank and the Child," *Global Social Policy* 10, no. 2 (2010): 172–92.
26 Williams, *Claims and Frames in the Making of Care Policies*, 12.
27 Thanidka Mkandawire, *Social Policy in a Development Context. SPD Programme Paper No.7* (Geneva: UNRISD, 2001); Michael Cichon and Woflgang Scholz, "Social Security, Social Impact and Economic Performance: A Farewell to Three Famous Myths," paper for the Dfid/GTZ/ILO Research Seminar, Social Security Department (Geneva: ILO, 2006).
28 Thandika Mkandawire, "Transformative Social Policy and Innovation in Developing Countries," *The European Journal of Development Research* 19, no. 1 (2007): 13–29.
29 Jenson, "Lost in Translation," 447.
30 Maxine Molyneux, *Change and Continuity in Social Protection in Latin America: Mothers at the Service of the State? GD Programme paper No.1* (Geneva: UNRISD, 2007).
31 Eleonor Faur, *The Care Diamond: Social Policy Regime, Care Policies and Programmes in Argentina* (Geneva: UNRISD, 2008).
32 Silke Staab and Roberto Gerhard, *Childcare Service Expansion in Chile and Mexico: For Women or Children or Both? GD Programme Paper No.10* (Geneva: UNRISD, 2010).
33 Paula England, Michelle Budig and Nancy Folbre, "Wages of Virtue: The Relative Pay of Care Work," *Social Problems* 49, no. 2 (2002): 455–73, Shahra Razavi and Silke Staab, Special Issue of *International Labour Review, Underpaid and Overworked: A Cross-National Perspective on Care Workers* 149, no. 4 (2010).
34 Rajni Palriwala and N. Neetha, "Care Practices and Care Bargains: Paid Domestic Workers and Anganwadi Workers in India," *International Labour Review* 149, no. 4 (2010): 511–28.
35 Marilyn Waring, interviewed by Cathy Cavanaugh, January 1998. http://aurora.icaap.org/talks/waring.htm.
36 World Bank, *World Development Report 2006: Equity and Development* (Washington DC: World Bank, 2006); World Bank, *Engendering Development Through Gender Equality in Rights, Resources and Voice. Policy Research Report* (Washington, DC: World Bank, 2001).
37 Diane Elson, "Gender Equality and Economic Growth in the World Bank World Development Report 2006," *Feminist Economics* 15, no. 3 (2009): 39.
38 Barbara Harriss-White, "Work and Well-being in Informal Economies: The Regulative Roles of Institutions of Identity and the State," *World Development* 38, no. 2 (2010): 170–83.
39 Harriss-White, "Work and Well-being in Informal Economies," 178.

13 Economic governance and the regulation of intimacy in gender and development

Lessons from the World Bank's programming[1]

Kate Bedford

- **An overview of the World Bank's revised approach to economic governance and GAD**
- **Gender balance at the project level: PROFAM**
- **Conclusion: some broader lessons**

> Explicit recognition is important. We all know that what is unnamed is more likely to be unsupported, ignored and misunderstood.
>
> (Paul Hunt, United Nations Special Rapporteur on the right to the highest attainable level of health, speaking on the importance of naming sexual rights in UN work)[2]

> The temptation
> To take the precious things we have apart
> To see how they work
> Must be resisted
> For they'll never fit together again.
>
> (Billy Bragg, song writer, writing of a miserable intimate relationship)[3]

There is a long queue of people wanting to have conversations about sexuality and its relevance to the future work of the United Nations Entity for Gender Equality and the Empowerment of Women (UN Women). Many of those people want to focus on sexuality and development in relation to reproductive health concerns, or education.[4] Another strand of work has focused on sexual rights—to bodily integrity and freedom from violence; to abortion; to accurate and nonjudgmental information about sex; rights for sexual minorities, etc.[5] These are crucial conversations, and they require urgent action if UN Women is to succeed as a feminist policy site.

That said, one of the aims of this chapter is to echo those arguing that sexuality is relevant to people interested in political economy, not just to those working on health and education, rights, and sexual minorities.[6] Specifically, I aim to consider how changing forms of economic governance are related to changing norms about intimate interpersonal attachment in gender and development (GAD) policy. I will do that by focusing on the GAD initiatives of the World Bank, the world's largest and most influential development institution.

The argument I present here is in three parts. The first briefly explains how the Bank's approach to development and economic governance has changed since the early 1990s, and how this helped trigger a new GAD approach linked to the emergence of new norms around interpersonal attachment between men and women in loving partnerships. The second part provides an example of a project intervention that stemmed from this new approach to economic governance and intimacy, a family-strengthening initiative in Argentina called PROFAM. The third part looks at the lessons that could be learned from the Bank's experiences. I focus on both the possibilities opened up for feminist development work through these new links between restructured markets and restructured intimacies, and the critical questions that scholars and practitioners committed to improving the work of international institutions on gender might want to ask about those links. This concluding part explores in some depth a key tension within current GAD scholarship and policymaking over the utility of disrupting commonsense norms about economic governance and intimacy. This tension is especially pertinent when gender policies have been given space in international institutions in part because they seem to shore up those norms, as when gender equality initiatives promise to generate more efficient markets, or more caring couples. My discussion exists in the midst of the two quotations that open this chapter, and it aims to dwell in that space for a while to see what lessons, if any, emerge as we think about the approach to women's economic and intimate empowerment that might be taken by UN Women. In this regard I very much take the point made by the book's editors that the new agency marks a crossroads for internationalist feminist politics, wherein the common senses we have mainstreamed can be fruitfully revisited.[7]

An overview of the World Bank's revised approach to economic governance and GAD

The World Bank—founded as part of the post-war Bretton Woods agreements—is a multilateral actor, with immense power to set the international development agenda in terms of its economic resources, its research activities, and its discursive power. In the 1980s and early 1990s, that agenda was a firmly neoliberal one, focused on policies of privatization, austerity in social spending, trade liberalization, and rapid capital market opening. Its development loans often came attached with multiple conditions, such as privatizing state assets or introducing user fees for public services.

However, in the 1990s the Bank moved away from this orthodox neoliberal approach. The institution began to place more emphasis on building the institutions necessary for free markets to flourish; on balance between states, markets, and civil society; on good governance; and on social development concerns such as inclusion of the marginalized and poverty eradication. This approach was championed by James Wolfensohn, Bank president between 1995 and 2005. Wolfensohn was determined to humanize the institution by broadening its understanding of development to include more than growth, and to give more attention to poverty, environmental sustainability, HIV/AIDS, youth, anti-corruption, empowerment, and so on. Wolfensohn was controversially replaced in 2005 by Paul Wolfowitz, chief architect of the US war on Iraq, but in his short tenure Wolfowitz did not change the Bank's new approach to development as holistic, as requiring balance between states, markets, and civil societies, and as requiring a focus on poverty, empowerment, inclusion and institutional strengthening. Robert Zoellick, the current Bank president, has continued these policies as well. In a recent speech announcing increased spending to support the Millennium Development Goals, he said: "We must … redouble efforts to target support to the poor and vulnerable. … And as we do, we always need to keep in mind that this work is ultimately about empowering people."[8]

There are many ways of talking about this formulation of capitalism. Geographers Peck and Tickell refer to it as inclusive neoliberalism. They distinguish between the 1980s process of "roll back" neoliberalism, involving the active destruction and discrediting of Keynesian welfarist and social collectivist institutions, and the current phase of "roll out" neoliberalism that tries to secure a deeper, more sustainable social transformation.[9] Porter and Craig refer to attempts to embed Third Way variants of capitalism, to make neoliberalism more socially sustainable and less prone to crises.[10] However, in the Bank the shift is referred to in terms of a move from a Washington Consensus (WC) to a Post-Washington Consensus (PWC);[11] hence that is the term I deploy here.

Crucially, the PWC entailed a new approach to economic governance. Conditionality was officially abandoned; more emphasis was placed on getting the private sector and civil society groups involved in development debates; and attempts were made in many countries to secure greater domestic ownership of reforms. For example Wolfensohn launched a tripartite civil society-government-World Bank review of structural adjustment policies in six countries in 1996, and considerable work went into public relations campaigns emphasizing the pro-poor nature of structural reform initiatives. Rather than seeing economic reform as something that could be imposed from above, the Bank shifted to seeing it as requiring interventions in numerous sectors, involving multiple stakeholders and extensive participation of nongovernmental actors. In such a framework, governance increasingly occurs through partnerships and networks involving not just formal political actors, but also civil society groups, communities, and private corporations. The objective of governance is also broadened in scope, aiming at a coordinated societal transformation in formal and informal institutions, laws, and attitudes at multiple levels. As Wolfensohn explained in a 1999 address:

> There can be no doubt that strong institutions at the local level are the real key to effective poverty reduction. It takes more than changing formal rules to build these institutions. It means changing the informal rules and norms. It means building people, building values, and building skills and incentives that can support peoples committed to change.[12]

There has been considerable debate about the significance of these changes to the Bank's development approach, but I focus here on the fact that the shift to a PWC in development was significant for gender policy, for a number of reasons. Most obviously, Wolfensohn's embrace of a less starkly free-market vision of development involved a new attention to gender concerns. He devoted more money and staff to gender issues, and he was publically supportive of gender empowerment initiatives. Moreover, the organization changed the direction of its gender policy in line with critique of orthodox neoliberalism. Since the 1980s the Bank had been criticized by feminists for the crisis of social reproduction engendered by neoliberal reform.[13] Social reproduction refers to the labor required to reproduce the social. It is usually unpaid and usually done by women. It includes biological reproduction, the reproduction of labor power, and social practices connected to caring, socialization and the fulfillment of human needs such as childcare, housework, subsistence agriculture, cooking, and voluntary work to sustain community organizations.[14] Under the WC, services providing support to social reproduction were cut at the same time that women were increasingly required to work for pay due to rising poverty and male unemployment.

Hence women were overburdened. In effect structural adjustment policies represented an attempt to (re)privatize social reproduction responsibilities on the backs of women,[15] assuming, to use economist Diane Elson's wonderful phrase, that their time was "infinitely elastic."[16]

In the Bank's shift to a PWC, this approach to gender—basically one of exhausting women—has changed. Emphasis on strengthening informal institutions, crisis-proofing, inclusion, building a pro-poor coalition for reform, and empowerment has played out in gendered ways, and has transformed the manner in which GAD is understood to relate to the Bank's overall economic development approach. Most obviously, one can see links between the PWC and gender in the fact that the family—understood narrowly as a privatized couple in a nuclear unit—has been positioned as a crucial informal institution necessary for good development. For example the 2000–01 *World Development Report* asserted the need to strengthen ties connecting family members as part of its effort to improve institutions and hereby reduce poverty.[17] Moreover, family is listed first in the Bank's online inventory of sources of social capital, on the grounds that "as the main source of economic and social welfare for its members, the family is the first building block in the generation of social capital for the larger society."[18]

In addition, while the Bank continues to argue that employment is key to women's empowerment (in line with its broader advice that poverty is best resolved through labor market reform), the neglect of care work in GAD has been challenged. There are now frequent references in Bank gender discussions to the importance of unpaid care and the dangers of over-stretching women who provide it, and a solution of sorts has been embraced—that women should participate in paid employment, and that men should be more involved in care. Male inclusion, targeted on poor men in particular, has hence become a key feature of gender policy. Indeed, Bank gender staff refer often to a crisis in gender relations, particularly a crisis in masculinity in which men have been stripped of socially valued roles as breadwinners.[19] Their masculinity thus wounded, poor, unemployed men are an increasingly dangerous threat to women and children. Put another way, feminist debate about the crisis in social reproduction stemming from structural reform has been rendered legible to the Bank as a debate about the crisis in the family, particularly evident in relation to poor men's alleged irresponsibility. The problem to which GAD policy is directed is thus a problem of men and women failing to partner up properly, and the solution is to increase the emphasis on partnership.

This emphasis on partnership is evident throughout the organization's development work, from the commissioning of research to the graphics of happy couples used in policy papers, from the funding of projects to the rhetoric about families used in Bank leadership speeches. For example, in a 2006 Bank-hosted conference on the Millennium Development Goal of gender equity, Wolfowitz gave an opening speech containing a segment entitled *A Cart with Two Wheels*, explained thus: "During my visit to Pakistan this past summer, one poor woman told me that development is like a cart with two wheels—one man and one woman. If one of the wheels isn't moving, the cart won't go very far."[20] The Bank's gender web page illustrated this conference with a graphic of a set of scales balanced by a male and female figure on each side—a perfect visual representation of how central male-female partnerships are to the world's largest development institution in its PWC phase.

Gender balance at the project level: PROFAM

This approach to GAD, aiming to promote a two-partner, sharing model of love and labor in which women are emancipated through employment and men are included in the family as responsible fathers and husbands, has been operationalized in a number of project and policy sites. To give a sense of what is involved here, of the spaces created for feminist policy entrepreneurship, and of the constraints operative within these spaces, in this section I turn to a Bank project to promote family strengthening in Argentina, called PROFAM. I conducted 34 interviews with groups and individuals associated with this project in 2007, and I visited subprojects that had received money.

Argentina is a key case study of the shift from a Washington to a Post-Washington Consensus approach to development, having embraced one of the most stringent economic reform packages in Latin America during the 1990s only to spectacularly fall from grace in the crisis of 2001.[21] The country enacted the largest default on external debt in modern economic history (so far), and over 50 percent of its population—once famed as the most middle class in Latin America—were plunged into poverty. This crisis had a particularly significant impact on the Bank, since it was (is) held partly responsible for the collapse.[22] The PWC imperatives to generate domestic ownership of reforms were particularly compelling in this case, as were the pressures on the Bank to show that it was addressing issues of participation, inclusion, and empowerment.

Bank gender staff and domestic gender advocates seeking Bank funding saw their scope for intervention in the Bank's Argentine policies wax and wane across these shifts in development approach. However, they have had to consistently contend with a difficult domestic environment, particularly around reproductive rights. In the late 1990s Argentina lacked any national reproductive health policy, information on contraception was unevenly accessible, and unsafe abortion was—and still is—the leading cause of maternal mortality.[23] When the Bank reached out to feminist civil society organizations in the 1990s for advice on where to target social lending as part of the PWC shift, reproductive health was identified as a key priority. It was an issue which the government, allied to the Catholic Church in this respect, did not support, so feminists turned to the Bank instead.

The strategically titled Family Strengthening and Social Capital Promotion Project, PROFAM (2001–6), emerged out of this context. This was a US $6.72 million initiative involving a $5 million loan from the bank, administered by Argentina's National Women's Council (CNM). PROFAM aimed to facilitate the launch of gender equity and family strengthening initiatives involving local governments, nongovernmental organizations (NGOs), and community organizations[24]—the multilevel governance imperatives of the PWC are very clear here. Its most important component was a grant fund given to local organizations for work on gender issues, but it also aimed to operationalize social capital concerns by strengthening poor households' internal links, and creating more bridging links to other households and organizations. The project was understood as a way to "strengthen family cohesion [and] solidarity between male and female members" of households, to "test the hypothesis that the promotion of a more cohesive and less segregated family life will positively impact the ability of families to face challenges posed by poverty."[25] Its key performance indicators included: "improved awareness among beneficiary families about the importance of gender equity and cohesive family practices as means of reducing their vulnerability," a decrease in domestic violence, and "improved evidence of responsible parenthood among beneficiary families … through: i) increase in father's involvement in childcare and support; and ii) decrease in teenage pregnancy."[26] Staff also claimed that PROFAM would provide economic benefits, by supporting heads of household affected by high unemployment adapt to new roles, and improving income-generating opportunities for women.[27]

In design terms, PROFAM therefore epitomizes the Bank's adjusted approach to gender policy as about generating family unity to prevent poverty and, to use the project's language, improve the resilience of the poor. It was being written up within the Bank as a best practices model of family capacity building when it had barely started, and it is referred to in multiple regional Bank documents as a model which "would provide unemployed men with new skills in fathering and care giving and make it easier for their female partners to seek work opportunities and excel on the job."[28] For these reasons, although it is a small loan PROFAM provides a crucial site in which to analyze what the new approach to GAD meant in *implementation*, and to study the alliances generated in practice.

In implementation terms, there was a range of projects funded with PROFAM money, including ones that were clearly counter-normative. These included many reproductive health initiatives, a project to empower sex workers facing police violence, and a project funding support services for survivors of domestic violence. The loan also created space for a range of organizations to interact with, and learn from, each another. One woman working in a Church project with the mothers of young drug addicts described what happened when she went with her colleagues to a PROFAM meeting in 2006, where NGOs that had received money were brought together to talk about their experiences:

> I went with two other women in the project. There was a table where two girls were sat, we asked if we could sit with them, and they said yes. They introduced themselves and one of them said to me "[her name]" and offered me her hand, and said "sex worker." And I said to her "Oh, so you are a social worker?" "No, no, a *sex* worker."
> (original emphasis, author's translation)

She was laughing at herself as she relayed this interaction with her table-mates, but she closed by saying, very seriously, that she had learned the following from it: "Definitely, we spoke of the same thing. Of the marginalization of women, of the space that women simply don't have. Really, we spoke of the same thing. It was very nice."

That said, however, this openness to various women's projects became less and less evident as the project progressed. In particular, once the economic crisis occurred the wide range of meanings attached to the term "family strengthening"—a range that included sex workers and (self-described) Church ladies—narrowed, in ways that raise important concerns. Firstly, one key consequence of a family strengthening/gender sharing framework was to grant the Catholic Church considerable power in GAD projects. To reiterate, the idea for PROFAM originated with Argentine feminist groups in the late 1990s. They turned to the Bank's gender team for support for reproductive health work because the state-level was blocked due to the opposition of the Catholic Church. Hence several of the early projects involved condom distribution, and awareness-raising about sexual health. Indeed the term "family strengthening" was adopted precisely in order to preempt opposition from the

Catholic Church, and early plans for PROFAM were run past Cáritas (the main vehicle for Catholic welfare services in Argentina) before being submitted to the government, to ensure that this key Church welfare agency had no objections.

Many feminist policymakers involved in the early stages, in and outside the Bank, felt that this use of Church language was a savvy strategic move, because family strengthening was so poorly defined that it could provide a cover for sex workers and domestic violence activists while disarming the Church. Some feminist NGOs said that they just manipulated the project's language to do what they wanted to do anyway, such as teach women their rights to prosecute abusers or distribute condoms.

The 2001/2 economic and political crisis altered the environment within which these strategies played out, however, changing the uneasy equilibrium between progressive and conservative visions of family strengthening in place in the design phase, and giving the Catholic Church far more power in the project than it had at the start. After the 2001 economic meltdown, the Church was given a formal role in the state's emergency response, and it got directly involved in the development of social plans.[29] Moreover, in 2003, in a conciliatory move towards conservatives, President Nestor Kirchner gave control of the CNM to María Lucila ("Pimpi") Colombo, a woman with strong links to conservative activists in the Catholic Church.[30] She had opposed the reproductive rights legislation brought to the floor when she was a Buenos Aires senator. As a result, Church-based organizations were far more prominent in the second call for projects (when Colombo was running the CNM) than in the first. For example, the housewife's union SACRA (linked to the Church and in which Colombo had long been active) got money from PROFAM for a microenterprise project in which women sewed baby clothes.

I note these shifts not to impugn projects carried out by Church-affiliated organizations, many of which have played a key role in providing services to the poor, and which affirm that they were nonjudgmental in their approach to family. Rather, I note them because the increasing involvement of such organizations in PROFAM was clearly related to the new approach to economic governance evident after Argentina's crisis. The Church got more power within PROFAM because it was given more power over social policy as a result of the crisis, and this caused considerable anger among some self-identified feminist groups involved at the start.

Secondly, after the 2001 meltdown PROFAM was reoriented as part of a broader redefinition of gender policy in Argentina as about restoring family unity in response to economic crisis. In 2006 the Argentine government described its development approach as focused on repairing and rebuilding the social fabric,[31] and the CNM's mission was redesigned and refocused on the need to strengthen the role of the family. As the head of social development in Argentina put it, when speaking about PROFAM: "Within this work in our communities, the key is the family, of this there is no doubt … If one wants to change reality, one has to work together with the family, *together* with the family, *with* the family."[32] This was a national project linked to the revival of Argentine values, which, it was argued, had been hurt after years of neoliberalism.

Coupled harmony and family regeneration were crucial components of that revival, as is clearly evident in the projects chosen to represent PROFAM. For example, the DVDs put out to celebrate and publicize the initiative repeatedly featured a couple, Angelina and Lorenzo. Filmed working on their garden plot together, they were a posterchild for the loan's success. One NGO, which stated that "our focus on gender should be seen more than anything as part of our work with the family, how the role of women and men plays out within the family," used PROFAM money to take families on recreational outings in order to restore their unity. The only example given in the DVD on the long-term results of PROFAM highlighted a project on family meals. In the DVD, the local coordinator explains:

> One of the things that we realized when we made the visit (to the community) … is that the table as an object had disappeared from the homes; a decade of community kitchens meant that the mother did not cook, that the table was not used as a communal space, where family and social and community problems were discussed … So what we proposed was how to go back to eating as a family.[33]

The project ensured that women got access to fresh vegetables, so they could cook nutritious meals for their families. State-funded community kitchens were closed, changed into family gardens. In this way, the project aimed "to recuperate the family table, as a space of dignity."

Some of these initiatives were criticized by organizations who had received money in the first round of funding for providing inadequate solutions to poverty and for further overburdening women.[34] However, PROFAM did, to some extent, attempt to grapple with this problem. After all, concerns about women's work burden and the need for the family to draw together to face adversity were central to the loan from the start. Hence several of the projects involving youth aimed at restoring family harmony by teaching young people to support women's involvement in income-generating work. One community organization interviewed described its workshop activities as involving asking:

> How do the man and the children cope with this new woman's role? ... The workshops had children doing learning activities that permitted them to understand and help their mother. So that you didn't end up with a battle ground in the family because the mother wants to go to work.

In contrast there was no systematic effort made within PROFAM to provide childcare services for women working in microenterprises. Better-shared family care was the preferred solution to the social reproduction tensions evident in the project, not nurseries.

Thirdly, PROFAM helped shape future GAD work via its uneven impact on different women's groups. Generally, established NGOs that got money through this project continued to find funding, although not necessarily through the Bank or the state—PROFAM did not seem to have affected their strength or weakness in any significant manner. However, some organizations that were less well-established at the start of the project got a considerable boost through PROFAM, if their activities easily fit its family strengthening, male-female partnership frame. These organizations have since received new funding through much larger social development interventions using the principles of family restoration in the aftermath of crisis pioneered in this loan. For example, a provincial NGO which received PROFAM money for an HIV/AIDS project and capacity-building workshops on gender subsequently received money from a provincial anti-drugs program, producing leaflets to raise awareness of drug abuse and the role of families in its prevention. These warned that drug abuse was linked to chaotic and disorganized family life, and the abdication of family responsibility to care for children. The self-described Church ladies—who said they had never received international funding before PROFAM and who, like many others, struggled valiantly with the Bank's complex accounting requirements to administer their small project—are now involved in the Ministry of Social Development's Programa Familias por la Inclusión Social (Families for Social Inclusion Program).[35] In contrast I was told that the sex workers union with which they sat and talked at the PROFAM meeting was defunct.

Conclusion: some broader lessons

I do not wish to over-generalize from PROFAM, and Argentina's gender policy environment in the immediate post-crisis period was obviously a very specific one. I also recognize that the Bank's model of GAD is distinctive, and evolving.[36] However the project, and the new approach to GAD and economic governance that it represents, is hardly an outlier in current GAD conversations. Some European Union welfare states have also moved from breadwinner-housewife assumptions about social reproduction to dual earner/caregiver models, where sharing partnership is a key concern of policymakers.[37] The Bank is not the *only* institution focusing on embedding neoliberalism in part by (re)privatizing responsibility for social reproduction within reconfigured families where women work more, men care better, and poor kin draw together to increase their resilience.

With this in mind I would like to close by suggesting three lessons that feminists might take away from the World Bank's experience of new approaches to economic governance and the regulation of intimacy. The first is the importance of remaining curious about, and critical of, *all* the levels of governance being brought together in new policymaking approaches. Theoretically, gender scholars tend to share a commitment to seeing levels of governance as social constructions, rather than seeing the level as pre-determined, natural, and/or outside of politics. The family and the community are almost always theorized as sites of power and politics by feminists, for example. PROFAM suggests that it is *especially* important to extend critical scrutiny to the family at present, given its central role in PWC policy frameworks. This example shows that transnational actors have become involved in attempts to govern interpersonal relations in a range of ways, including via family strengthening activities and efforts to increase sharing within heterosexual couples. Levels of governance are interacting in extremely complex ways here, with multilateral lenders seeking to bolster certain intimate attachments as part of new, more inclusive development agendas, and acting in alliance with state actors, feminists, and religious actors. It is not simply that the Bank is dumping responsibilities downward, onto families that are always-already configured; rather, the production and regulation of certain kinds of interpersonal intimate attachments has increasingly become part of the Bank's business. Our understanding of what responsibilities rightly belong to the family, versus other levels, and of what conjugal attachments should mean politically and economically is thus shifting as the family becomes increasingly relevant to, and the object of, governance.

Secondly, the Bank's experiences in PROFAM confirm the value of a multi-pronged approach not just to who is doing the governing, but also to what is the objective of the governance. There is no way to understand the emergence of the gender-sharing approach within the

Bank (or to reveal its limits) without taking seriously its *dual* status as a model of economic governance, and as a model of governing intimacy. As I hope I have shown, gender sharing in the Bank is very much about shifts in political economy, particularly shifts in social reproduction. It focuses on the adjustment of intimate labor burdens within loving couples as a solution to the problems posed by the WC, and in this way it is very much linked to debates about the political economy of care, and the links between globalization and changing norms around sexuality. Hence we can only understand the Bank's mobilization of a new regime of normative heterosexuality if we link it to shifts in macroeconomic policy. However gender sharing is not *just* about market efficiency, or ensuring that families are functional for capitalism (to use an older language). It also rests on key assumptions about normative gender and sexuality—and a key lesson here is that these assumptions need to be spoken about, debated, and at times contested.

Thirdly, and finally, PROFAM suggests the urgency of staging that debate now: it is already past due. The costs and benefits of gender sharing as a project of economic and intimate governance are becoming increasingly apparent. For example, analysis of PROFAM shows that the space opened by the project for transgressive use was precarious and short-term, and feminist NGOs were not necessarily the most significant beneficiaries of the intervention. However, the appeal that the gender sharing approach holds for feminists is also clear, whether instrumentally (because it helps disarm potential opponents of gender equality) or normatively (because feminists themselves cherish the ideals being promoted). The companionate ideal of balance at work in PROFAM also provides space for feminist seizure, in that it takes on board critique of the WC for overburdening women and ignoring unpaid care, while moving beyond nostalgia for breadwinner-housewife solutions to social reproduction tensions.

However, while gender advocates may have wanted international financial institutions to take care more seriously, we did not necessarily want childcare erased as a policy priority, replaced by more shared (but still privatized) caring labor within loving couples. We may not agree that family cohesion is—or should be—a poverty-eradication strategy. We may not want poor men held responsible for women's poverty. And some of us are deeply concerned that our conversations about a crisis in *social reproduction* have been rendered legible to global governing institutions as attempts to prevent a crisis in the *family*. There are, then, dangers in the approach, and *they are already evident*. Put briefly, it can privilege a limited model of women's empowerment that hinges solely on their formal labor force participation (a source of disquiet to many feminists who do not necessarily see employment as the source of women's liberation); it can bolster religious actors whose approach to gender equality is contested by many feminists; it can aid in the privatization of social reproduction; it can target already marginalized groups of men as particularly irresponsible and uncaring; and it can limit our ability to talk about intimate autonomy and sexual freedom outside of a male-female coupled model.

At present there is little debate about those dangers; the consensus story about good gender governance is so successful that it can be hard to see its limitations, still less imagine and build for an alternative. Moreover, some feminists fear that talking about the limitations might puncture the consensus. Taking it apart to see how it works might be dangerous, because we might never get it back together again and we might cause a divorce—from both the World Bank's macroeconomists, and the Church. Being explicit may also harm the under-the-radar spaces available to those who have smuggled in condoms and sex workers to family strengthening projects.

My own preference is to take that risk: to advocate for an explicit conversation about who gains, who loses, and what becomes harder to fight for in women's organizing when we endorse this ideal of governing markets and intimacies through sharing balance. But given that others have such different views, I would suggest that whatever we think of the Bank there is immense value, for ourselves, of a frank conversation about what *we* understand by economic and intimate empowerment, and the role—if any—that we want transnational institutions to take in regulating those spheres. It is my hope that if we can better understand the consequences of how markets and sexualities are currently entangled in international governance projects, we can better decide what we prefer as alternatives.

Notes

1 This chapter is adapted from Kate Bedford, *Developing Partnerships: Gender, Sexuality, and the Reformed World Bank* (Minneapolis, MN: University of Minnesota Press, 2009). It is reprinted with permission from the publisher.
2 Cited in Andrea Cornwall, Sonia Corrêa and Susie Jolly, ed. *Development with a Body: Sexuality, Human Rights and Development* (London: Zed Books, 2008), xi.
3 Billy Bragg, "Must I Paint you a Picture?" *Workers Playtime* (London: Go! Discs, 1988).
4 Cornwall, Corrêa and Jolly, ed., *Development with a Body.*
5 Alice Miller, *Sexuality and Human Rights* (Versoix, Switzerland: International Council on Human Rights Policy, 2009); Amy Lind, ed., *Development, Sexual Rights and Global Governance* (New York: Routledge, 2010).

6 Substantial work has been done on links between economic and sexual regulation. For example Jacqui Alexander was arguing over 15 years ago that structural adjustment relied on policies to induce normative heterosexuality among the poor, a point honed by US analysts of welfare reform. See M. Jacqui Alexander, "Not Just (Any) Body Can Be a Citizen: The Politics of Law, Sexuality and Postcoloniality in Trinidad and Tobago and the Bahamas," *Feminist Review* 48 (1994): 5–23; Cathy Cohen, "Punks, Bulldaggers, and Welfare Queens," *Gay and Lesbian Quarterly* 3 (1997): 437–65.

7 Gülay Caglar, Elisabeth Prügl and Susanne Zwingel, "Introduction," this volume.

8 *World Bank Pledges New Agriculture, Education, and Health Spending to Help Poor Countries Achieve Millennium Development Goals by 2015* (World Bank document, press release number: 2011/087/HDN), 13 September 2010, http://web.worldbank.org/WBSITE/EXTERNAL/NEWS/0,contentMDK:22699316~pagePK:64257043~piPK:437376~theSitePK:4607,00.html.

9 Jamie Peck and Adam Tickell, "Neoliberalizing Space," *Antipode* 34, no.3 (2002): 380.

10 Doug Porter and David Craig, "The Third Way and the Third World: Poverty Reduction and Social Inclusion in the Rise of 'Inclusive' Neoliberalism," *Review of International Political Economy* 11, no.2 (2004): 387–423.

11 E.g. Shahid Javed Burki and Guillermo Perry, *Beyond the Washington Consensus: Institutions Matter* (Washington, DC: World Bank, 1998).

12 *Voice for the World's Poor: Selected Speeches and Writings of World Bank President James D. Wolfensohn 1995–2005* (Washington, DC: World Bank, 2005), 161–62.

13 Pamela Sparr, ed., *Mortgaging Women's Lives: Feminist Critiques of Structural Adjustment* (London: Zed Books, 1994).

14 Isabella Bakker and Stephen Gill, ed., *Power, Production and Social Reproduction: Human In/Security in the Global Political Economy* (Toronto: York University, 2003).

15 Janine Brodie, "Shifting the Boundaries: Gender and the Politics of Restructuring," in *The Strategic Silence: Gender and Economic Policy*, ed. Isabella Baaker (Totowa, NJ: Zed Books, 1994), 46–60.

16 Diane Elson, "Gender-Aware Analysis and Development Economics," *The Political Economy of Development and Underdevelopment*, sixth edition, ed. Kenneth P. Jameson and Charles K. Wilber (New York: McGraw Hill, 1996), 71.

17 World Bank, *World Development Report* (Washington, DC: World Bank, 2001), 128.

18 World Bank 2009, *Social Capital*, http://go.worldbank.org/VEN7OUW280.

19 See especially Ian Bannon and Maria Correia, ed., *The Other Half of Gender: Men's Issues in Development* (Washington, DC: World Bank, 2006).

20 Paul Wolfowtiz 2006, *Opening Remarks at High-Level Consultation: Promoting the Gender Equality MDG: The Implementation Challenge*, www.worldbank.org/gender/.

21 Argentina became a model adherent of the WC under Carlos Menem (1989–99) and his Finance Minister Domingo Cavallo. The country's economic collapse in 2001 thus devastated faith in neoliberalism, and discredited the political elite. See Judith Teichman, *The Politics of Freeing Markets in LA: Chile, Argentina and Mexico* (Chapel Hill, NC: University

of North Carolina Press, 2001); Steven Levitsky and María Victoria Murillo, ed., *Argentine Democracy: the Politics of Institutional Weakness* (University Park, PA: Penn State University Press, 2005).

22 María Alejandra Corbalán, *El Banco Mundial: intervención y disciplinamiento: el caso Argentino, enseñanza para América Latina* (Buenos Aires: Editorial Biblos, 2002); Susana Murillo *et al.*, ed., *Banco Mundial: Estado, Mercado y Sujetos en las Nuevas Estrategias Frente a la Cuestión Social* (Buenos Aires: Centro Cultural de la Cooperación Floreal Gorini, 2006).

23 Sandra Cesilini and Natalia Gherardi, ed., *Los Límites de la Ley: la salud reproductiva en la Argentina* (Buenos Aires: Banco Mundial, 2002).

24 World Bank, *Project Appraisal Document for PROFAM*, 30 October 2001 (Washington, DC: World Bank), 8.

25 World Bank, *PROFAM Project Information Document*, 31 July 2000 (Washington, DC: World Bank), 1–3.

26 World Bank, *Project Appraisal Document for PROFAM*, 2.

27 Ibid., 7–10.

28 Maria Valéria Pena and Maria Correia, *World Bank Brazil Gender Review: Issues and Recommendations* (Washington, DC: World Bank, 2002), 45.

29 Martín Dinatale, *El Festival de la Pobreza: el uso político de planes sociales en la Argentina* (Buenos Aires: Konrad-Adenauer-Stiftung, 2004), 186.

30 Débora Lopreite, *Gender and Welfare Regime Change in Argentina: International Conventions and Feminist Networks in a New Social Agenda* (paper presented to Annual Meeting of Canadian Political Science Association: Toronto, 2 June 2006).

31 World Bank, *Country Assistance Strategy for the Argentine Republic 2006–8* (Washington, DC: World Bank, 2006), 72.

32 PROFAM, DVD 2 of 9 DVDs on PROFAM (CNM Buenos Aires: CNM, 2006), original vocal emphasis.

33 PROFAM, DVD 2 of 9 DVDs on PROFAM (CNM Buenos Aires: CNM, 2006).

34 This has long been a criticism of Argentina's government-supported poverty assistance programs. See Javier Auyero, *Poor People's Politics: Peronist Survival Networks and the Legacy of Evita* (London: Duke University Press, 2001); Dinatale, *El Festival de la Pobreza*; Constanza Tabbush, *Contesting Gender Narratives in Development Policies: Women and Conditional Cash Transfers in Argentina*, Women and International Development Working Paper #292 (East Lansing, MI: Center for Gender in Global Context, Michigan State University, 2008).

35 This is also known as Plan Familias. For analysis see Tabbush, *Contesting Gender Narratives in Development Policies.*

36 For example, on the distinctive way in which gender is treated in the International Finance Corporation, see Kate Bedford, "Doing Business with the Ladies: Gender, Legal Reform, and Entrepreneurship in the International Finance Corporation," *Labour, Capital and Society* 42: 1& 2 (2009), http://www.lcs-tcs.com/PDFs/42_12/Abstracts/Bedford.pdf.

37 See Gillian Pascall and Jane Lewis, "Emerging Gender Regimes and Policies for Gender Equality in a Wider Europe," *Journal of Social Policy* 33, no. 3 (2004): 373–94.

14 Feminist strategies and social learning in international economic governance

Gülay Caglar

- **Feminist strategies in international economic governance**
- **The concept of learning and policy change**
- **Normative knowledge and social learning in international economic governance**
- **Conclusion**

> There is nothing so disempowering as the idea that there is no alternative.[1]

Feminist activism in the field of international economic governance has grown out of an emphatic rejection of the rhetoric that "There Is No Alternative" to neo-liberal macroeconomic policies promoted by international financial institutions (IFIs) and the World Trade Organization (WTO). Feminists point to the gendered effects of those policies as a result of deep-seated gender biases in mainstream economics and, as a consequence, in economic policy-making. In order to overcome such gender biases both in international discourses and decision-making processes, transnational feminist networks specifically pushed for a gender-equitable approach to macroeconomic policy-making as a meaningful alternative to neo-liberal policy designs. An example for such an approach is, for instance, the idea of gender budgeting: a policy tool which implies the examination of fiscal policies—that is, the broader fiscal strategy (Keynesian or Monetarist) that governments pursue to stimulate economic growth as well as the composition of public expenditures and revenues—from a gender perspective.[2] Gender budgeting helps to question core macroeconomic assumptions and policy priorities. Ultimately, it is a tool to identify ways of how to reprioritize fiscal policies towards greater gender equality. Feminist activities in this area predominantly center on international economic institutions as these institutions, with their policy prescriptions, are setting the frame for economic policy-making at national and local levels. In this context, activists have applied a strategy of knowledge-based gender advocacy that I will call "knowledge networking," that is they generate and diffuse knowledge about the interrelationship between gender and macroeconomic policies by networking with feminist economists. Such networking is of utmost importance as feminist economists provide the expertise on how to realize gender-equitable macroeconomic policies.[3]

In the past ten years, the strategy of knowledge-based gender advocacy has turned out to be relatively successful as claims for gender-equitable macroeconomic policies have gained considerable attention in international discourse. In fact, the idea of gender budgeting has even been brought into the operational agenda of several international organizations.[4] Most projects seeking to advance gender-equitable macroeconomic policies are concerned with initiating and/or supporting gender budget initiatives in various countries of the Global South by providing funds as well as expertise on gender budgeting (for example training manuals). This has led to the spread of gender budget initiatives all over the world, which can be deemed a successful outcome of feminist interventions in the field of international economic governance. However, many incidents show that the seemingly successful feminist agenda-setting did not culminate in significant policy change. Gender budgeting has rarely been applied as a policy tool to change a government's fiscal strategy or taxation policies. Moreover, reactions to the economic downturn in the wake of the subprime crisis have revealed that the idea of gender-equitable economic governance in general and gender budgeting in particular has not been fully internalized by decision-makers at international and national levels. For example, economic stimulus packages that were promptly enacted by many governments to overcome short-term, demand-side problems have been largely male-biased. These packages have included measures that are primarily directed towards male-dominated industries (i.e., automobile or construction).[5] This has been the case also in countries, such as South Africa, that have used gender budgets. What have been left out of consideration are the fiscal consequences of these measures in the long run: Increasing budget deficits in the course of crisis resolution have negative effects on future policies—massive retrenchments in expenditures particularly in the area of public social services are required in many countries. This predominantly affects women, as worldwide experiences with economic restructuring in the past two decades have shown.

International organizations are not as gender blind as national governments. They are aware of the gendered impacts of the crisis, yet they draw different conclusions: While, for example, the World Bank regards "women as agents of change"[6] and accordingly puts emphasis on advancing women's economic empowerment, the United Nations Development Programme (UNDP) makes reference to the idea of gender budgeting, however, without directly linking crisis resolution policies to a reprioritization of macroeconomic goals.[7] That means, the World Bank aims at enabling women to participate in the market economy (either as entrepreneurs or as employees) and the UNDP focuses on protecting women from the worst impacts of the crisis. Thus, despite the successful agenda-setting of transnational feminist networks, gender issues are either completely absent in macroeconomic discourses and policies or addressed in ways that do not question core macroeconomic assumptions and priorities.

Why is this the case? Why does policy change not take place? Drawing on the literature of social learning in policy processes I argue that policy change is closely linked to the extent to which international economic institutions learn from knowledge-based gender advocacy. Accordingly, in a first step I ask how exactly feminists have brought gender issues into the field of international economic governance and how international organizations have learned from them. Reflecting the concept of social learning from a Foucaultian perspective, I further argue that learning is not an uncontested process but imbued with interpretative struggles about what actually the problem is. In that context, I focus—as Carol Lee Bacchi suggests—on "problem representations."[8] Whether or not a situation is perceived as problematic depends on which meaning is discursively assigned to it. Consequently, in a second step I show that the extent to which international organizations learn depends on how they interpret the problem of gender in macroeconomic policy-making. I illustrate that international organizations assign different meanings to gender budgeting. The analysis will show that learning is just taking place within the specific interpretative framework or power/knowledge-nexus, respectively, operating in international organizations.

Feminist strategies in international economic governance

The type of strategy (bottom-up or top-down) that is pursued by feminist actors varies according to which political arena and which economic policy (sub)fields it relates. The three main strategies employed have been lobbying in intergovernmental meetings, the institutional strategy of gender mainstreaming, and influencing the framing of economic policy making by discursive politics and knowledge-based advocacy: Bottom-up strategies, like lobbying, are particularly pursued during intergovernmental policy processes, such as during WTO Ministerial Conferences or during United Nations (UN) conferences, like the UN Conference on Financing for Development (UN-FfD) in order to integrate gender equality concerns into international negotiations and agreements. Feminist activists directly lobby country delegations and policy-makers trying both to influence the wording of documents and to attain commitments to undertake concrete measures towards gender equality. The participatory space officially provided by the UN-system allows for the exertion of influence on intergovernmental meetings. Women's organizations with consultative status with the UN Economic and Social Council (ECOSOC) are eligible to take part in UN meetings and conferences.[9] In contrast, there is no comparable formal participatory mechanism within the WTO. Consultative status with the WTO is given in a very selective manner, which is why the field of international trade policy can be characterized as highly anti-democratic.[10] Therefore, women's groups dealing with international trade issues, like for example the International Gender and Trade Network (IGTN), deploy an "inside-outside" strategy; that is an information sharing strategy between those activists who are accredited and work from the inside and those without consultative status that concentrate on mobilization from the outside (through tactics such as awareness raising or protest actions).[11]

Intergovernmental processes, such as the WTO Ministerial Conference in Seattle in 1999 and the UN-FfD conference, prompted feminists to mobilize extensively in the field of international economic governance. Particularly the UN-FfD conference with its preparatory process was extremely important for feminist activists, as this conference provided a forum for deliberating on issues of international economic governance at UN-level. In fact, this conference was unique in two ways: First, the negotiation on ways of financing development included "systemic issues"—that is, the role of the international monetary, financial and trading systems for development. Second, the UN-FfD conference introduced the approach of multi-stakeholder consultations, which included heads of state, ministers, and representatives of civil society, private-sector business and of international organizations, such as the World Bank, International Monetary Fund (IMF) and WTO. This setting offered an exceptional opportunity for feminists directly to get in dialogue with representatives of international organizations that are normally not easily accessible. In the course of the UN-FfD process feminist activists were able to integrate the idea of gender budgeting into the final document—the so-called Monterrey Consensus.[12]

For many feminists, though, the results of the conference were not far-reaching enough. One point of criticism is that the conference failed to challenge seriously the existing neo-liberal framework. Another one is that gender equality concerns were not incorporated throughout all key areas, including especially systemic issues.[13] To give an example: The Monterrey Consensus urges to "[s]trengthen the coordination of the United Nations system and all other multilateral financial, trade and development institutions to support economic growth, poverty eradication and sustainable development worldwide,"[14] however, without explicitly mentioning that economic growth, poverty eradication and sustainable development shall be facilitated in a gender-equitable manner. Despite these flaws, concessions articulated in the Monterrey Consensus are notable for providing opportunities for feminist activists to hold national governments and international agencies accountable for what they have agreed to.

Yet another set of feminist strategies, like for example gender mainstreaming (see Jacqui True and Laura Parisi in this volume), aims at the incorporation of gender equality concerns into all policies and programs of international organizations, regional agencies and national governments. Gender budgeting is an important element of gender mainstreaming as it is an instrument to ensure that all commitments made to gender equality are resourced adequately. Furthermore, gender budgeting itself is a mainstreaming strategy in the field of macroeconomic policy-making as it aims at systematically integrating gender into budgets—local and national budgets as well as those of international organizations—and fiscal policies.

Together with feminist economists, activists intensively advocated for gender budgeting at the international level. This, generally, indicates a positive stance towards gender mainstreaming and institutional processes. In the context of international trade, however, the strategy of gender mainstreaming is much more contested, which acutely became evident in 2003 at the Fifth WTO Ministerial Meeting in Cancun, Mexico: While some women's organizations (e.g., Millennium Feminista) advocated instituting a "gender committee" within the WTO, others (e.g., IGTN) emphatically argued against a mainstreaming strategy within the multilateral trading system.[15] Their refusal of gender mainstreaming was a consequence of their fundamental criticism of the WTO in its current (undemocratic and inequitable) format—its rule-making, review and enforcement mechanisms, its ideology and its political agenda, which does not include any commitments to gender equality and women's human rights (see Mariama Williams in this volume). They refused to mainstream gender into a "wrong" system. As Maria Riley puts it: "If the goal of gender mainstreaming is to integrate women into the existing economic, political and social structures so that women can 'get their share' it lacks the critical lens that feminist groups have been bringing to the debate on social, economic and political transformation."[16] Evidence from the Asia-Pacific Economic Cooperation (APEC) shows that the adoption of gender mainstreaming has not changed the regional trading regime to a gender perspective. It has rather, as Jacqui True argues, "quieted the criticism about […] gendered economic structures in the export-led economies of East Asia."[17] This is exactly the reason why IGTN pursues the goal to transform the current neo-liberal economic regime and not to get gender incorporated into it.

Finally, two overarching strategies which support the strategies of lobbying and gender mainstreaming are discursive politics and knowledge-based advocacy or knowledge-networking, respectively. Discursive politics is "the effort to reinterpret, reformulate, rethink, and rewrite the norms and practices"[18] in international economic governance. Thus, discursive politics is "the politics of meaning making."[19] Driven by the rejection of the gender-blind and neo-liberal economic discourse, feminist activists in collaboration with feminist economists seek to reframe economic debates. The goal is to disrupt the mantra about "sound" macroeconomic policies as these policies are solely judged on market-oriented criteria disregarding non-marketed activities, such as social reproduction which heavily rests on women's unpaid work. Feminists argue that allegedly prudent fiscal and monetary policies with their one-sided focus on productivity and efficiency in the formal economy lead to an overburdening of the care economy at household level; the costs of efficiency in the market economy are shifted to the non-market sector and eventually to women. Against this backdrop, feminists intend to discursively unpack the meaning of "sound" macroeconomic policies and, thereby, to shift the focus on the "social content"[20] of macroeconomic policies. Ultimately, they strive for the acknowledgement of the care economy and women's unpaid work, respectively, as an integral part of the formal economy.

For these reasons activists as well as gender representatives of international organizations (e.g., the UNDP) collaborate with feminist economists, who provide expertise in regard to gender issues in macroeconomic policy-making. This expertise goes beyond being simply technical in character; feminist economists do not just impart their knowledge about *how* to "engender" macroeconomic policies but also provide the factual as well as ideational reasoning on *why* to realize gender-equitable macroeconomic policies.

One particular network of feminist economists, the "International Working Group on Gender, Macroeconomics and International Economics" (GEM-IWG), has been particularly influential at the international level. GEM-IWG was formed in 1994 by feminist economists who identify themselves as heterodox, that is, they distance themselves from orthodox neoclassical economics. The aim of the network is both academic and political in character: On the one hand, its members seek to challenge the mainstream within the economic discipline by developing gender-aware macroeconomic models and theories. GEM-IWG provides the space for alternative knowledge creation and its dissemination. On the other hand, the objective is to contribute to a policy change by translating their theoretical insights into political action and by disseminating alternative policy ideas, like gender budgeting, internationally. For these purposes the network established the so-called "Knowledge Networking Program on Gender, Macroeconomics and International Economics" in 2003 partly funded by the Ford Foundation and partly by the UN Division for the Advancement of Women (DAW). This program is a capacity-building program targeted at economists in academia, governments (working in the Central Bank or Ministry of Finance), non-governmental organizations (NGOs) and international organizations.[21] The program focuses on training key multipliers who then disseminate their newly acquired knowledge within their institutional contexts. Representatives from women's organizations, for instance, organize so-called "economic literacy" programs for activists at the local level who are not trained in economics at all. These kinds of capacity-building and economic literacy programs have contributed to a significant mobilization of feminist activists in the field of international economic governance. In fact, many transnational feminist networks, such as the Women's International Coalition for Economic Justice (WICEJ), Women in Development Europe (WIDE) or the International Gender and Trade Network (IGTN), base their analyses and advocacy work on the knowledge produced by the members of GEM-IWG.

To build up knowledge on gender and macroeconomic policy-making is pivotal for feminist interventions in the "hard" area of economic policy-making: It enables activists and gender advocates to: a) define what exactly the problem is; b) use the correct economic terminology; and c) precisely formulate their claims. Moreover, feminist claims that are grounded in economic knowledge are more difficult to be dismissed by policy-makers and bureaucrats within international organizations. Feminists are often accused of basing their claims on anecdotal evidence. Being aware of that, feminist economists are eager to build up their arguments on formal proofs. That means they obey the rules of knowledge generation in the economics discipline and put emphasis on integrating gender into formal macroeconomic modeling. As one member of the network states:

> I think our group [GEM-IWG] was important by bringing a level of *formalism* and a level of rigour or *mathematical rigour* to a set of issues that feminists have been discussing for many, many years. And I think that it's not to downgrade or denigrate in any way the work that feminist economists have done. It's just putting it into a different format that has made it recognized in a different way.
>
> (Interview GEM-IWG I, 18.06.2003)

Thus, there is a consciousness that feminist claims are taken more seriously if they refer to the same frame of reference as that of economic policy-makers. This, however, involves the danger that feminists subject themselves to the practices of knowledge generation and thereby reinforce and reproduce the epistemic hegemony of the economics discipline.

Irrespective of the ambiguity of being heard by policy-makers, the knowledge network GEM-IWG has been an amplifier for feminist interventions in the field of international economic governance as it has built up and strengthened strategic partnerships between scholars, women's NGOs and feminist bureaucrats in that field. Particularly the strong links to UN institutions, like the former UNDAW and UN Development Fund for Women (UNIFEM), have significantly contributed to spreading the idea of gender budgeting.

Spreading an idea, however, does not necessarily go hand in hand with a policy change as already indicated above. Although the idea of gender budgeting has been taken up by several international organizations, macroeconomic policies basically remain neo-liberal and gender blind. Thus, the question emerges, *how* international organizations adopt the idea of gender-sensitive fiscal policies without actually changing their principal policy priorities. What do international organizations learn from knowledge-based gender advocacy? What impedes learning?

The concept of learning and policy change

Learning is an important aspect that has to be considered when analyzing processes of policy formulation and implementation in international governance. International organizations constantly learn in one or another way; they adjust their policy goals, strategies and policy instruments in reaction to past experiences and new knowledge.[22] Thus, learning is a process that ideally culminates in policy change.

Scholars of policy studies differentiate between different types of learning in order to grasp the degree to which actors alter their political behavior.[23] Peter Hall, who has developed one of the most sophisticated theories of learning and policy change, refers to the notion of social learning as an important dimension of policy-making.[24] Hall distinguishes three levels of changes vis-à-vis three central components of policy-making: the overarching policy goals, policy instruments and the precise settings of these instruments. First order change (incremental change) is the process by which instrument settings (e.g., the level of social expenditure) are modified, while the basic policy instruments and the overarching goals remain the same. Second order change occurs when policy instruments as well as the settings are altered. Third order change takes place when policy-makers additionally define a new hierarchy of goals in policy-making. This implies radical changes in the policy paradigm—a paradigm shift. To sum up, first and second order change are cases of what Peter Hall calls normal policy-making, that is "a process that adjusts policy without challenging the overall terms of a given policy paradigm."[25] In contrast, third order change entails modifications in all three components of policy-making.

According to Hall, the pressure for a paradigm shift arises when the existing policy paradigm is not able to provide comprehensible explanations and solutions for a specific policy problem, as he shows in his seminal work on macroeconomic policy-making in Britain. Hall identifies three factors that facilitate a paradigm shift; a new paradigm has to be economically, administratively and politically viable in order to prevail. That means, new ideas have to: 1) convince experts of the economics profession; 2) "fit the long-standing administrative biases of the relevant decision makers and the existing capacity of the state to implement them"[26]; and, 3) correspond with the goals and interests of the dominant political actors. Thus, in his analysis, institutional aspects as well as power politics are the main explanatory factors for third order change. Hall pays astonishingly little attention to the role of knowledge, although he emphasizes that the process of learning is based on new information or knowledge. He conceptualizes the relationship between knowledge and politics as a cause-effect relation, which is shaped by the broader institutional framework. Accordingly, institutions ultimately constrain or enable the influence of a specific paradigm, and thus determine policy outcomes. Yet, what remains unclear is why a specific policy problem is perceived as a problem and not potentially another one. Hence interpretative struggles over the correct problem definition, and as a result, over the *meaning* of a specific policy paradigm are not theorized. This, however, is important if we want to understand how and under which circumstances social learning and policy change takes place or does not.

In this respect, Peter May identifies an important dimension of learning, that is social policy learning, which "entails lessons about the social construction of policy problems, the scope of policy, or policy goals."[27]

Adding a Foucaultian perspective to this dimension of learning, I contend that knowledge plays a constitutive role: All interpretations of the policy problems are made against the background of a specific knowledge that draws the boundaries of acceptable behavior and approaches in a specific field of action. This means whether or not gender issues in macroeconomic policy-making are regarded as problematic depends on the organizational background knowledge about the interrelationship between gender and economic issues. This background knowledge is both scientific and normative in nature. I refer to "gender knowledge"[28] and normative economic knowledge in international organizations in order to explain how different meanings are assigned to gender budgeting.

I argue that a paradigm shift in international economic governance takes place if international organizations are able to learn lessons about the construction of the problem that has to be solved in regard to gender and macroeconomic policy-making. Or to put it differently: Although international organizations have put the goal of gender budgeting on their operative agendas, a paradigm shift has not occurred, as they have not changed their interpretations of the policy problem. International organizations have altered the setting of their policy instruments and at best also their policy instruments (first and second order change), for example by introducing tools and methods for the analysis of public expenditures from a gender perspective. However, their interpretations of the problem in regard to gender issues in economic policy-making remain the same, as will be shown below.

In the following section, I focus on the gender knowledge underlying the organizational rationale of the World Bank and UNDP. I have chosen these two organizations as they are the most influential in the international economic development discourse. In addition to their annual flagship publications, that is the *World Development Report* (World Bank) and the *Human Development Report* (UNDP), I look at the way they frame issues of gender and economic development in their manuals, strategy papers, and similar publications.

Normative knowledge and social learning in international economic governance[29]

Both the UNDP and the World Bank support activities on gender-sensitive expenditures throughout the world. At first glance their engagement seems to be based on common sense and on similar assumptions. However, upon closer examination it becomes clear that there are considerable differences in what these organizations are targeting through gender budgeting.

In order to understand why different meanings are assigned to gender budgeting, we need to identify how the organizations interpret the interrelationship between gender and macroeconomic issues. They interpret the problem in regard to gender and macroeconomic policies by referring to cause-effect relations—either they stress the effects of macroeconomic policies on women (how do macroeconomic policies impact women's lives?) or they analyze the effects of gender inequalities on the outcome of macroeconomic policies (how do gender inequalities impede the smooth functioning of the market?). I refer to the notion of "storylines" which has been developed by scholars of interpretative policy analysis. Storylines are defined as "a sort of narrative that allows actors to draw upon various discursive categories to give meaning to specific physical or social phenomena."[30] They "condense large amounts of factual information intermixed with the normative assumptions and value orientations that assign meaning to them."[31]

Two "storylines" can be identified through which international organizations construct the object of political intervention: If emphasis is put on the effects of macroeconomic policies on women (of the macro on the micro level), as in the case of the UNDP, the collective situation of women and the care economy is put at the center of gender budgeting. In that case, all policy measures are directed towards ameliorating the living conditions of women. Thus, "engendering" or gender budgeting, respectively, encompasses social compensatory measures, which mitigate the negative effects of macroeconomic policies. In this context, it is argued that the care economy has to be recognized as an essential part of the formal economy. The principal aim of gender budgeting is to support the care economy by implementing a few social policies measures that are particularly targeted at women's needs. If, however, emphasis is put on the effects of gender inequalities at the household or institutional level for the outcome of macroeconomic policies (of the micro/meso on the macro level), as in the case of the World Bank, gender budgeting embraces another meaning: Discriminatory practices within society as well as the traditional gendered division of labor are regarded as barriers for women actively to participate in the market economy. It is argued that these hindrances for women imply market distortions, and thus, inefficiency in the economy because a significant part of the labor force is not mobile. Consequently, the focus of "engendering" lies on overcoming the barriers that women face at household or institutional levels. This implies the goal of liberating women from their reproductive responsibilities and from societal marginalization. Policy measures such as educational programs target women's participation in the market economy—either as employees or as entrepreneurs. In these two storylines the economic sphere is abstracted from the social sphere: The gender perspective in the first storyline directs attention to the *social* component of economic policies, whereas the gender perspective in the second storyline highlights the *economic* value of gender equality.

The analysis of UNDP documents reveals that UNDP policies are based on a gender knowledge favoring a traditional gender division of labor, in which principally women are responsible for unpaid care work. The care economy is regarded as the emotional and social sphere which stabilizes the society. The UNDP emphasizes women's important role for maintaining families and argues that their "vital social functions […] become only too visible when juvenile delinquency rates rise, the elderly are left to die alone or cultural traditions wither."[32] The care economy is deemed to be the core of the social fabric, which is fundamental for human development. As the UNDP stresses, "[w]ithout genuine care and nurturing, children cannot develop capabilities, and adults have a hard time maintaining or expanding theirs."[33] Further, UNDP-policies are based on economic knowledge that is principally critical to economic development; the UNDP strictly differentiates human development from economic development. The latter, according to the UNDP, is confined to economic growth, which does not automatically lead to the welfare of human beings. Thus, economic growth is considered a necessary, but not sufficient component of human development. The UNDP attempts to broaden the notion of development to non-economic issues, which are issues of physical, psychological and social well-being. In this sense, the care economy is deemed an integral part of the economy, and it is argued that women's unpaid work has to be valued. Nonetheless, the boundaries between the social and economic spheres are drawn by simply adding the social to the economic. This becomes evident if one closely examines what the UNDP means by the valuation of women's work. It advocates for the valuation of women's reproductive work, however, "without advocating an exchange value for all non-monetized activities and without arguing for a radical shift in the way families organize their work."[34] That means, assuming that women are better at nurturing children and caring for family members, the aim is to maintain the traditional division of labor in order to enhance human development. It is on the grounds of this normative knowledge base that "engendering" is constructed as a *social* issue. Hence, "engendering" means that women's reproductive work has to be valued by implementing a few social compensatory measures that support women's unpaid work at the household level. Social policy measures are simply added to economic policies.

The World Bank also directs attention to women as *mothers* and states that "[t]he influence of a mother on her child begins in the womb and continues through preschool and later childhood […] In critical ways she determines her child's early intellectual stimulation and physical development."[35] In contrast to the UNDP, the World Bank simultaneously emphasizes the importance that women be healthy, well-educated and income-earning mothers. The underlying assumption is that mothers are much more concerned about the intellectual and physical development of their children than fathers are. Accordingly, the World Bank expects that women are willing to spend part of their own income for the well-being of their families. The World Bank refers to studies in which "[t]heir ability to save and invest in their families is well documented."[36] For the World Bank, investing in children is pivotal as they are deemed the "future workforce"[37] of the economy. In this way women are regarded as an "engine for growth and development."[38]

Evidently, the World Bank is primarily concerned with the smooth functioning of the economy. All constraints hindering the free flow of market forces have to be overcome. Within this reasoning, the World Bank emphasizes the importance of investing in infrastructure facilities (e.g., medical care), which enable women to care *efficiently* for their children. According to the World Bank, "improvements in infrastructure […] directly affect the efficiency of home production, reducing time spent in household work and releasing time for other activities."[39] Thus, the objective is to increase the productivity of mothers in household work and their participation in the market economy. The World Bank constructs women's identities as caring *and* working mothers. Thus, women are deemed economic subjects; however, not in the sense of *homo oeconomicus* as women in the first place are presumed to be altruistic. The gender knowledge of the World Bank corresponds to an economic knowledge that is based on micro-foundational theories (such as human capital theory or new institutional economics): Mothers who are able to invest in their children directly contribute to human capital accumulation and, thus, to economic growth. Within this ideational context, a totally different meaning is assigned to the object of gender budgeting: It is regarded as an (economic) *investment strategy in mothers* for growth. This is why gender budget exercises promoted by the World Bank predominantly focus on educational, health and income-generating measures.

To sum up the results of the analysis: Within the UNDP the problem that has to be solved is interpreted against the background of a gender knowledge which can be characterized as *traditional* in the sense that it emphasizes women's reproductive role in the care economy. Moreover, within the UNDP, the notion of the economic is imbued with a negative connotation; the UNDP dissociates itself from policy approaches that aim to achieve development through accelerating economic growth in the formal economy. Thus, the normative economic knowledge of the UNDP is characterized by a rejection of economic policy approaches that put emphasis on economic development without considering human development as an end in itself. From this view, the boundaries between social and economic phenomena are drawn by repeatedly emphasizing the value of the social sphere, and thereby accepting the separation between the social and the economic. Consequently, the meaning of gender budgeting is fixed by defining gender issues as social issues.

In the case of the World Bank the emphasis lies on liberating the economy from societal barriers such as gender divisions of labor. It is not the care economy as such that is put at the center, but women as mothers; however, in contrast to the UNDP, women here are income-earning, healthy and efficient mothers. Hence, women are regarded as both mothers *and* economic subjects. Accordingly, the objective is to free women from their reproductive responsibilities and to endow them with skills for the labor market. Both gender knowledge and economic knowledge in the World Bank are characterized by a positive stance towards the market economy. From this perspective, gender budgeting is regarded as an instrument of economic governance, which primarily focuses on expanding educational and income-generating services. Gender budgeting is grasped as a policy solution that is functional to the economy. Thus, the boundaries between the social and the economic spheres are drawn by subjecting social issues to the logics of economic growth.

Conclusion

Feminist actors employ a variety of strategies in order to bring gender issues into the field of international economic governance; they lobby governments and international organizations, they advocate for gender mainstreaming and they engage in discursive politics. In this context, gender expertise or knowledge is pivotal as it determines whether feminist interventions are taken seriously or not. That means knowledge-based gender advocacy is an important strategy in the field of international economic governance—gender-sensitive macroeconomic policies are made plausible through scientific evidence on the interrelationship between gender and macroeconomics. This is an area in which UN Women should focus its activities: A transformation of economic policy-making must begin from a change in economic orthodoxies that keep social and economic issues separate.

As the analysis shows, there are considerable differences on how international organizations refer to gender issues in the context of economic governance. The type of gender politics in the field of international economic governance varies depending on how gender is put in relation to social and economic issues. In this respect, normative knowledge—or gender knowledge—plays a constitutive role. Ultimately, gender knowledge in connection with economic knowledge determines which meaning is assigned to the object of political intervention (gender budgeting) and thereby demarcates whether gender issues are regarded as a social or economic problem. In other words, normative knowledge constitutes the object of political intervention by constructing the policy problem that has to be solved. This normative knowledge, once metamorphosed in organization-specific world views and practices, is highly institutionalized and, consequently, durable and not easy to alter. This is why normative knowledge poses a serious cognitive as well as institutional impediment to social learning and to a paradigmatic policy change within international organizations. The idea of gender-sensitive expenditures has been adopted by international organizations for the purpose of either investing in women as mothers (World Bank) or of compensating for women's social disadvantages (UNDP). Core macroeconomic assumptions and policies, however, remain. Consequently, in order to challenge macroeconomic discourses feminists need to direct their attention towards the social construction of policy problems and the normative knowledge base of international organizations. This requires a reflexive discursive strategy, that is critical reflections in regard to (own) practices of knowledge generation and the interrelationship between objectified academic knowledge and normative knowledge. The new UN Women's agency could serve as both a catalyst and pioneer in this effort.

Notes

1 Diane Elson and Nilüfer Cagatay, "The Social Content of Macroeconomic Policies," *World Development* 28, no. 7 (2000): 1360.

2 Diane Elson, "Integrating Gender Issues into National Budgetary Policies and Procedures: Some Policy Options," *Journal of International Development* 10, no. 7 (1998): 929–41.

3 The term "gender-equitable" is usually used in the context of macroeconomic policies and denotes the aim of ensuring fair conditions for women and men in the economy through which gender equality is ultimately achieved.

4 For example, of the Commonwealth Secretariat, the World Bank and the United Nations Development Programme (UNDP), which closely collaborated with the United Nations Fund for Women (UNIFEM). However, the World Bank does not use the term "gender budgeting." The activities of the Bank in this area appear under the title "Economic Policy & Gender:

Public Expenditure." See World Bank 2011, *Economic Policy & Gender: Public Expenditure*, http://go.worldbank.org/ZKKS51GEQ0.

5 Germany, for instance, passed two huge stimulus packages in 2008/09 which comprise a total amount of approximately €80 billion. One of the most striking measures of the stimulus packages was the so-called "car scrapping scheme" (*Abwrackprämie*) which was designed to encourage consumers to scrap vehicles that were at least nine years old and to buy a new car. The government provided a subsidy of €2,500 per car and spent €5 billion in total. See Organisation for Economic Co-operation and Development (OECD), *OECD Economic Surveys: Germany 2010* (Paris: OECD, 2010).

6 World Bank 2011, *In the Midst of the Financial Crisis, Women Can Be Agents of Change*, http://go.worldbank.org/RQC7937NT0.

7 Anna Fälth and Mohammad Abbadi, *UNDP Guidance Note Gender Equality and Poverty Reduction: Financial and Economic Crisis*, no. 1 (2009), http://content.undp.org/go/cms-service/stream/asset/?asset_id=3231022.

8 Carol Lee Bacchi, *Women, Policy and Politics: The Construction of Policy Problems* (London: Sage Publications, 1999); Carol Lee Bacchi, *Analysing Policy: What's the Problem Represented To Be* (Frenchs Forest: Pearson Education, 2009).

9 Peggy Antrobus, *The Global Women's Movement. Origins, Issues and Strategies* (London: Zed Books, 2004).

10 Dera J. Liebowitz, "Governing Globalization: Feminist Engagements with International Trade Policy," in *Global Governance. Feminist Perspectives*, ed. Shirin S. Rai and Georgina Walyen (Houndsmill, NY: Palgrave Macmillan, 2008), 207–33.

11 International Gender and Trade Network (IGTN), "IGTN's Activities in Cancun," *IGTN Monthly Bulletin* 3, no. 5 (2003): 7, http://www.igtn.org/pdfs/191_July03bulletin.pdf.

12 United Nations (UN), *Monterrey Consensus of the International Conference on Financing for Development. The Final Text of Agreements and Commitments Adopted at the International Conference on Financing for Development Monterrey, Mexico, 18–22 March 2002* (document A/CONF.198/11) (United Nations Department of Public Information, 2003), http://www.un.org/esa/ffd/monterrey/MonterreyConsensus.pdf, para 19.

13 Isabella Bakker, *Key Policy Initiatives on Financing for Gender Equality and the Empowerment of Women, Written Statement* for Commission on the Status of Women, 52nd session, New York, 25 February–7 March 2008, http://www.un.org/womenwatch/daw/csw/csw52/panels/key-policy/Panel%20on%20Key%20policy%20on%20FFGE-%20Paper%20by%20BAKKER.pdf, 2.

14 UN, *Monterrey Consensus of the International Conference on Financing for Development*, para 64.

15 Maria Riley, "Gender in the Multilateral Trade System. The Ongoing Debate," *IGTN Monthly Bulletin* 3, no. 8 (2003): 12–13, http://www.igtn.org/pdfs/276_OctNovbulletin03.pdf.

16 Ibid., 13.

17 Jacqui True, "Gender Mainstreaming and Regional Trade Governance in Asia-Pacific Economic Cooperation (APEC)," in *Global Governance. Feminist Perspectives*, ed. Shirin S. Rai and Georgina Walyen (Houndsmill, NY: Palgrave Macmillan, 2008), 151.

18 Mary Fainsod Katzenstein, *Faithful and Fearless: Moving Feminist Protest inside the Church and Military* (Princeton, NJ: Princeton University Press, 1998), 17. On discursive politics see also Mary E. Hawkesworth, *Globalization and Feminist Activism* (Lanham, MD: Rowman & Littlefield Publishers, 2006); Emanuela Lombardo, Petra Meier and Mieke Verloo, ed., *The Discursive Politics of Gender Equality. Stretching, Bending and Policymaking* (New York: Routledge, 2009).
19 Katzenstein, *Faithful and Fearless*, 17.
20 Elson and Cagatay, "The Social Content of Macroeconomic Policies."
21 The network still organizes such capacity-building workshops, however, at regional and country-level. See http://www.gem-europe.org.
22 See definition of learning in Peter A. Hall, "Policy Paradigms, Social Learning, and the State: The Case of Economic Policymaking in Britain," *Comparative Politics* 25, no. 3 (1993): 275–96.
23 Hugh Heclo, *Modern Social Politics in Britain and Sweden* (New Haven, CT: Yale University Press, 1974); Paul A. Sabatier and Hank Jenkins-Smith, ed., *Policy Change and Learning: An Advocacy Coalition Approach* (Boulder, CO: Westview Press, 1993); Peter J. May, "Policy Learning and Failure," *Journal of Public Policy* 12, no. 4 (1992): 331–54; Ian Greener, "Social Learning and Macroeconomic Policy in Britain," *Journal of Public Policy* 21, no. 2 (2001): 133–52.
24 Hall, "Policy Paradigms, Social Learning, and the State," 275–96.
25 Ibid., 279.
26 Peter A. Hall, "Conclusion: The Politics of Keynesian Ideas," in *The Political Power of Economic Ideas. Keynesianism across Nations*, ed. Peter. A. Hall (Princeton, NJ: Princeton University Press, 1989), 371.
27 May, "Policy Learning and Failure," 337.
28 Gender knowledge can be defined as those kinds of knowledge "which exist on society regarding the difference between the sexes, the reasoning of its self-evidence and evidence, the dominant normative ideas about the correct gender relations and divisions of labor between men and women." Sünne Andresen and Irene Dölling, "Umbau des Geschlechter-Wissens von ReformakteurInnen durch Gender Mainstreaming?" in *Was bewirkt Gender Mainstreaming? Evaluierung durch Policy-Analysen*, ed. Ute Behning and Birgit Sauer (Frankfurt/M., New York: Campus, 2005), 175 (own translation).
29 This section draws on Gülay Caglar, "Multiple Meanings of Gender Budgeting: Gender Knowledge and Economic Knowledge in the World Bank and UNDP," in *Gender Knowledge and Knowledge Networks in International Political Economy*, ed. Brigitte Young and Christoph Scherrer (Baden-Baden: Nomos Verlagsgesellschaft, 2010), 55–74.
30 Maarten A. Hajer, *The Politics of Environmental Discourse: Ecological Modernization and the Policy Process* (Oxford: Oxford University Press, 1995), 56.
31 Frank Fischer, *Reframing Public Policy. Discursive Politics and Deliberative Practices* (New York: Oxford University Press, 2003), 87.
32 UNDP, *Human Development Report 1995* (Oxford: Oxford University Press, 1995), 98.
33 UNDP, *Human Development Report 1999* (Oxford: Oxford University Press, 1999), 77.
34 UNDP, *Human Development Report 1995*, 97.

35 World Bank, *Integrating Gender into the World Bank's Work. A Strategy for Action* (Washington, DC: World Bank, 2002), 78.
36 World Bank, *Sustainable Development in a Dynamic World. Transforming Institutions, Growth, and Quality of Life. World Development Report 2003* (New York: Oxford University Press, 2003), 71.
37 World Bank, *Integrating Gender into the World Bank's Work*, 83.
38 World Bank, *Sustainable Development in a Dynamic World*, 71.
39 World Bank, *Integrating Gender into the World Bank's Work*, 186.

15 Structural power and the gender biases of technocratic network governance in finance[1]

Brigitte Young

- **Structural power and the gender biases in finance**
- **Groupthink in global finance knowledge and policy networks**
- **The shift to private and semi-private market governance structures**
- **Financial governance without women in regulatory and supervisory networks**
- **Conclusion**

Financial market governance, by far the most influential and important economic policy field with a sizeable impact on everyday life, is conducted largely behind closed doors, excluding those who are most affected by the spill-over effects into other areas of the "real" economy. Its power to shape and determine the structure of the global economic system is immense. Particularly the control over credit is a form of structural power[2] affecting financial, economic, political and social outcomes of every nation and also of ordinary citizens around the globe. Those who are in the position to create credit will, as Susan Strange pointed out many years ago, control the capitalist economy.[3] At the same time, the creation of policy rules that limit the discretion of national policy makers with respect to monetary and financial sector policies, allows the banks, insurance companies, hedge funds and other financial intermediaries to move money around the world and create new financial instruments. These policy rules exercise a structural power in that they have inscribed in them a series of biases that have gender as well as class- and race-based outcomes.[4]

One of the most striking features of the prevailing financial governance architecture is the shift of the regulatory competence from public to private institutions. Far from assuming that finance is democratically accountable, it is shaped by public-private networks incorporated into multi-level institutional processes and practices at the global, regional, national and local levels. The prevailing financial system is organized in a manner to provide private goods for the benefit of the few. Yet financial stability as well as the provision of an infrastructure to finance investment and consumption opportunities are not private goods, but global public goods. The financial sector is thus of strategic public interest. While the shift to private institutions has attracted wide attention in political science,[5] the low or even non-existent female representation in financial market regulation has—apart from some media attention—met with little interest in academic research.[6]

This chapter argues that a gendered analysis of the nature of representation, risk and regulation is important for a deeper understanding of the finance sector. Whether the financial crisis of 2007–9 could have been avoided if more women had been at the helm of large investment banks is a hotly contested issue. But the crisis has turned a spotlight on the low or even nonexistent female representation in top positions of financial institutions and in the decision-making bodies in key regulatory institutions, central banks, formal and informal financial networks. The same applies to today's efforts at global regulatory reform that excludes those who are affected most by the financial crisis: workers, minorities, women and representatives from the civil society, and less developed countries. Any regulatory reform should be built on a normative view stipulating that the task of finance is to serve society in terms of growth, efficiency and equity. A gendered approach is important, since it focuses on factors such as rights, equitable outcomes, and norms that value collective over private interests. Thus the exclusion of women raises questions not just about who sits at the table of financial governance, but who benefits and why, what are the social benefits of the present financial system, what are the risks of the increasing financialization of everyday life, what are the particular gendered biases inscribed in the financial governance structure, and most importantly, who makes the rules. The exclusion of women from finance is thus also a question of representation and democratic governance. In a time of increasing financial risks, women are called upon to provide the ultimate social safety net for their families when all other forms have failed, but at the same time they are excluded from changing macroeconomic policies that address questions of economic justice. The ultimate goal of financial institutions is not to operate according to financial criteria. Equally important should be social criteria which need to be based on a different understanding about macroeconomic policies. The major task for heterodox economists is to overcome the silencing of voices which try to challenge the gender biases in finance.

My intent in pointing out the underrepresentation of women is not to advance an essentialist view of the crisis as inherently male. Rather, the knowledge base that underpins the financial sector, the private knowledge network structures which reinforce class and gender hierarchy, and the pursuit of wealth which excludes broader social groups, all conspire to and create a climate for groupthink. Increasing female representation in decision-making bodies is a necessary but not sufficient precondition to safeguard specific female interests. Female representation only makes a difference if it comes with an alternative concept of financial order. However, while economic analyses of gender-related issues have increased in scope and rigor in fields such as labor market economics, social and fiscal policies, financial market economics and regulation have remained an exclusive male terrain.

The chapter proceeds as follows. Two major principles are introduced to shed light on the largely unanswered question why women are absent in the higher echelons of finance. The first section introduces the concept of the structural power of finance in determining outcomes, and of gender biases inscribed in these structures. As Susan Strange had pointed out, an essential aspect of structural power is the power over ideas. Whoever is able to define and control the channels by which a specific kind of finance knowledge is communicated exercises a very special kind of structural power.[7] The second part focuses on the knowledge-based networks of financial experts with their highly technical and specialized expertise, which can be characterized as exclusionary club arrangements. Diane Stone[8] argues that the dissemination of financial knowledge and ideas are closely related to exclusionary clubs equipped with an authoritative claim to policy-relevant knowledge, which is largely technocratic giving precedence to a particular conception of financial knowledge. These exclusionary networks operate largely to the disadvantage of women and reinforce gender inequality. Before citing some empirical examples of the absence of women in top decision-making structures of key public and private regulatory institutions and networks in the fourth part of the paper, the third section discusses the importance of some of the private and semi-private governance structures in finance which shape and determine the contemporary neoliberal economy. The conclusion speculates on whether the financial crisis of 2007–9 has opened a policy space for heterodox finance knowledge. It also provides some recommendations for actions to place more women into decision-making in the financial governance structures.

Structural power and the gender biases in finance

The present market-based financial system has evolved as a result of aggressively implementing regulatory changes promoting Anglo-Saxon governance modes of self-regulation. These changes have led to higher income and wealth inequalities by weakening unionized industrial relations, privatizing pension schemes and making the principle of social solidarity less acceptable in society.[9] Financial governance modes furthermore have a decisive impact on the degree of financial stability. The Anglo-Saxon regulatory model, which was pushed to the fore by the financial industry with strong support of the vast majority of mainstream economists, is a representation of structural power to shape the international political economy. Susan Strange has defined structural power in contrast to relational power as the power to "shape and determine the structures of the global political economy within which other states, their political institutions, their economic enterprises, and (not least) their scientists and other professional people have to operate."[10] Structural power derives its power from the control over ideas and knowledge and from the ability to deny access to others who hold different views.

Neoclassical economics in general—and traditional financial economics in particular—have assumed that markets are populated by disembodied rational actors pursuing profit maximization contributing to efficiency and overall welfare. In the early twenty-first century, finance as an academic discipline and professional practice is increasingly conceptualized as independent from arenas such as politics, society, and culture. The efficient market hypothesis has been the key proposition of neoclassical finance for almost 40 years.[11] Investors are assumed to be rational and hence value assets rationally. As a consequence asset prices are believed to incorporate all the available information immediately. But even if irrational investors enter the market, rational arbitrageurs are assumed to be cognizant of the mispricing and bring prices back to equilibrium levels. The doctrine of efficient markets in finance is first of all politically enforced by strong vested interests that have been able to capture the state. It is worth remembering Robert Cox's famous intervention that "Theory is always *for* someone and *for* some purpose."[12] Despite the assumption that the efficient market hypothesis is based on a principle of neutrality, there is no such thing as a theory in itself, as Cox reminds us, and if a theory presents itself as such, it is all the more important to question its ideology. States have increasingly formed alliances with private actors to push for financial liberalization since the 1980s. Both state actors and the private financial industry were able to steer globalization toward neoliberal outcomes. In this context, a discourse that reiterated neoliberal norms and praised the efficiency gains achieved through liberalization, privatization and regulation played an important role in foreclosing alternative knowledge of finance. While in the past mostly priests have exercised such structural power over governments, today the modeling techniques of financial experts decide over who has access to information and how this information is made available.

Financial knowledge entails a conceptual silence about gender,[13] and reflects a set of assumptions that fiscal and monetary policy and financial regulation, the key instruments of financial governance, are gender-neutral. The silence about gender is strategic in the sense that this silence obscures the ways in which financial governance interacts with, and reinforces, or restructures, the social organization of gender relations that prevail at different historical conjunctures. It is not enough to inquire into how women are represented in decision-making structures; it also is important to understand what tacit assumptions and biases about gender are part of scientific knowledge claims in finance.[14]

Diane Elson and Nilufer Cagatay have identified three such biases: the male breadwinner bias, the commodification bias, and the deflationary bias.[15] A male-breadwinner bias in macroeconomic policy regimes comes from assuming that integrating women into the labor market does not have to meet the criteria applied to a full-time breadwinner model. Instead, women are seen as secondary, mostly part-time workers, who only provide additional cash needs of the family. The deflationary bias refers to the monetary policies of giving priority to low inflation and fiscal restraint, which has a more negative impact on women, since governments are largely restricted through such policies to promote full employment and increases in health and education expenditures. Closely related is the third bias, the commodification bias, which occurs when public services are no longer provided through state-financed social services, but are provided increasingly through the market. This means that the work of social reproduction in less affluent families is absorbed into the unpaid sphere of families, and in the case of the more affluent this work is done through hiring of domestic, often migrant, workers.[16]

In addition to these three macroeconomic biases in policy regimes, there are two additional biases, which are mostly associated with the present global finance governance. These are: risk bias and creditor bias.[17] Financial governance has a decisive impact on how risk sharing is organized in society. With few if any savings and limited ownership of financial and real wealth, women are affected in a particular, negative way by the individualization of risk. The risk bias refers to downloading the risks of financial instability to those who can least afford it. Closely related is the creditor bias, which goes to the heart of the functions and institutions of the financial system with quite different distributional impacts for women and men. This bias results from the social relations between debtors and creditors. On one side of the money relationship is the claimant (monetary asset) and on the other the debtor (corresponding obligations). This relationship is highly conflictual, since the creditor owns and disposes over the monetary assets while the debtor has the monetary obligation imposed by the owners of money. The staggering amount of privatized debts of many women and minorities and the subsequent high rates of repayment problems, delinquencies, and foreclosures in the subprime crisis, which erupted in the United States in the summer of 2007, is a case in point. It demonstrates how market discipline helps to sustain the prevailing forms of dominance and domination.[18]

Groupthink in global finance knowledge and policy networks

The creation and dissemination of finance knowledge is no longer the sole domain of academia and the state, rather knowledge diffusion is largely done through various knowledge-based networks which are exclusive and exclusionary clubs equipped with an authoritative claim to financial knowledge. Knowledge networks are understood as producers, disseminators and users of a particular technical and specialized expertise, in the case of finance, "relying principally on scientific modeling, forecasting and risk-assessment as the basis for financial governance arrangements."[19] The emergence of knowledge and policy networks in global governance can be viewed as a characteristic feature of the new forms of global governance.[20] These networks have in many ways facilitated the international diffusion of neoliberal regulatory practices and their coordination across the globe. Globalizing elites are at the forefront of funding and staffing think tanks, foundations, and research institutions engaged in the production and dissemination of ideas. More importantly, these institutions impose a rationality that gives precedence to a particular conception of knowledge and is instrumental in restructuring the global economy. These international networks of members of the state bureaucracies, independent regulatory agencies, representatives of the financial industry and, most importantly, economic experts have wielded tremendous structural power in influencing global outcomes. This is part of a more general process of restructuring the state-market relationship on a global scale, in which increasingly financial market regulation is far less legitimated through the political process of nation-states. Instead, there is a growing role for the professional deployment of various kinds of expert knowledge. In parallel to the retreat of the state from financial regulation, political debates on such issues as the importance of regulation for systemic stability, its relevance for the corporate governance and for income and wealth distribution have become less important or even nonexistent in these networks.

A certain groupthink[21] has developed as a result of these networks whose members hold a common set of causal beliefs and share notions of validity based on internally defined criteria for evaluation, common policy objectives, and shared normative commitments. In regulatory authorities and bodies, research and analysis are motivated by the internal logic of self-referential epistemic communities employing mathematical models, rather than by a desire to understand how the financial markets work and their implication for the overall welfare, distribution and security. This governance model based on groupthink was forcefully advanced by—exclusively white male—efficient market financial economists. They viewed the emergence of new financial instruments and a shadow banking system that evolved outside the scope of existing regulation as efficiency enhancing, including the now infamous speculative investment vehicles, elements of the originate and distribute securitization process as well as new off-balance sheet financial vehicles established by banks.[22]

Alternative approaches with a broader horizon to include the insights of other social sciences and heterodox economics were mostly disregarded.[23] In the rare cases when women are allowed to enter the clubs, they have to adhere to the prevailing logic and beliefs defined by male traditional economists. Those who try to intervene and offer alternatives to the reductionist understanding of finance are either not employed, or if they are part of financial or regulatory institutions, they are marginalized and denied career opportunities. The most prominent example is the effort undertaken in the late 1990s by Brooksley Born, at the time head of the US Commodity Futures Trading Commission, to regulate derivatives markets. She incited fierce opposition from Alan Greenspan and then Treasury Secretary Robert Rubin. Michael Greenberger, who was a senior director at the commission, recalls that "Greenspan told Brooksley that she essentially didn't know what she was doing and she'd cause a financial crisis. Brooksley was this woman who was not playing tennis with these guys and not having lunch with these guys. There was a little bit of the feeling that this woman was not of Wall Street."[24] Born's recommendations were suppressed and she left government. Having the power to discredit suggestions to regulate derivatives demonstrates both relational power as well as structural power of Anglo-American finance to determine regulatory outcomes.

Before presenting some empirical examples of the exclusionary nature of the financial market regulatory networks, the next section will discuss the importance of some of the private and semi-private financial governance institutions and networks, which assert structural power to shape the contemporary global economy.

The shift to private and semi-private market governance structures

In global finance, the state has gone particularly far in accepting the shift in regulatory competence to private institutions, as indicated by the evolving mechanisms of self-regulation and co-regulation in financial governance. From an institutional perspective, two broad types of private actors can be identified. First, private autonomous authorities or institutions are involved in the establishment and enforcement of rules. Private credit rating agencies (CRAs) played an important role at various stages in the subprime crisis.[25] They assigned credit ratings for issuers of certain types of debt obligations as well as the debt instruments themselves and developed guidelines that are enforced by portfolio investors or bond issuers.[26] Under the Basel II agreement of the Basel Committee on Banking Supervision (BCBS), banking regulators were empowered to recommend credit ratings from certain approved CRAs.[27] The BCBS is an informal forum of supervisors and central bankers of the G10 countries, which formulates broad supervisory standards and guidelines and recommends statements of best practice in banking supervision. It is closely linked to the interests of large financial private institutions. Another important private institution with rule-setting authority is the International Accounting Standards Board (IASB) that develops the International Financial Reporting Standards and promotes the use and application of these standards. At a more informal level the International Institute of Finance (IIF), the world's only global association of financial institutions with important regulatory proficiencies, is also an important rule-setter. The IIF regularly issues principles of conduct that are supposed to be implemented by its members.

Second, public or semi-state institutions, such as independent central banks or supervisory authorities, create rules with binding or nonbinding character. Private actors bear the enforcement responsibilities. The BCBS, where the principle of self-regulation of banks is dominant, is a case in point. In other areas, autonomous private actors are involved in the rule-setting process. In the last decade this kind of public-private governance has gained importance with the establishment of the Forum for Financial Stability (FSF) in the late 1990s, which was set up after the Asian crisis on Germany's initiative. This body comprises representatives of the ministries of finance, central banks and supervisory bodies of the G7 countries, as well as other five major international finance centers. Private, semi-state or public international standard-setting bodies and other groupings are also members of the FSF. For example, representatives of the International Accounting Standards Board sit alongside governments in the FSF. The FSF develops indicators for early warning systems, conducts studies and identifies potential dangers for financial market stability. However, the recommendations it makes are not binding. In an effort to overhaul the international financial architecture, the 2009 Group of Twenty (G20) London summit agreed to establish a successor to the FSF, the Financial Stability Board, which includes all G20 countries, FSF members, Spain, and the European Commission.

While state-private fora are dominant in developing rules in financial governance, there is one rare example of public rule setting and public enforcement. This is the Financial Action Task Force on Money Laundering (FATF), which was founded by the Group of Seven (G7) in 1989. Its goal is to combat money laundering and terrorism financing. It is currently made up of 32 countries and territories and two regional organizations. However, the FATF works only on the basis of recommendations and has no legal means at its disposal to implement them.[28]

The next section will give some empirical examples of the absence of women in these private-public top decision making regulatory and supervisory organizations and networks (financial governance without women) in the European Union (EU) and at the global level. The exclusion of women and other civil actors has, as argued in the previous section, led to a single-minded groupthink which blinded its adherents to the increasing problems in the global financial sector prior to 2007.

Financial governance without women in regulatory and supervisory networks

If we look at the empirical evidence, we find that the female representation in the supervisory architecture of the EU is particularly low. Table 15.1 depicts the new supervisory system put in place in 2009 replacing the previous Lamfalussy architecture, which proved unable to ensure a stable financial sector.[29] The reformed structure consists of the European Systemic Risk Board (ESRB), which is responsible for macro-prudential supervision. It will be chaired by the president of the European Central Bank. In addition, three additional European Supervisory Authorities (ESAs) were created as the micro-prudential arm of EU-level supervision. These are the European Banking Authority (EBA), located in London, and responsible for the banking sector; the European Securities and Markets Authority (ESMA), located in Paris, responsible for the securities sector and financial markets; and the European Insurance and Occupational Pensions Authority (EIOPA), located in Frankfurt, and responsible for the insurance and occupational pension sector. These European supervisory authorities (ESAs) will each comprise about 40–50 people

Table 15.1 Female representation in reformed financial supervisory structures at the EU level

Institution	*Female*	*Number of members*	*Chair*
European Banking Authority (EBA)[1]	10%	32	M
European Insurance and Occupational Pensions Authority (EIOPA)[2]	7%	32	M
European Securities and Markets Authority (ESMA)[3]	22%	30	M
European Systemic Risk Board (ESRB)[4]	6%	37	M

Notes:

[1] European Banking Authority 2012, *Members and Observers*, http://www.eba.europa.eu/Aboutus/Organisation/Board-of-Supervisors/Members-and-Observers.aspx.

[2] European Insurance and Occupational Pensions Authority 2012, *Board of Supervisors*, https://eiopa.europa.eu/about-eiopa/organisation/management/board-of-supervisors/index.html.

[3] European Securities and Markets Authority 2012, *ESMA Board of Supervisors*, http://www.esma.europa.eu/page/esma-board-supervisors.

[4] European Systemic Risk Board 2012, *Organisation and Structure, General Board,* http://www.esrb.europa.eu/about/orga/board/html/index.en.html

in the first year, and about 100 after three years. Despite the reformed supervisory structure at the EU level one thing did not change: Males remain in the driving seat. The supervisory chairs of the three micro-prudential boards as well as of the ESRB are all males. The female representation on the Boards of Supervisors is also low. Even the 22 percent in the ESMA is below a critical mass of around 30 to 40 percent needed to break through the barriers silencing alternative views.

Table 15.2 gives an overview of female representation in select public as well as private global institutions with regulatory competence. It is interesting that only after the former director of the International Monetary Fund (IMF), Dominique Strauss-Kahn, experienced legal problems in the United States in 2011, did the office of director pass to a woman, Christine Lagarde, the former French finance minister. The lack of women in decision-making bodies in intergovernmental institutions is well documented.[30] Less work has been done on the private-sector institutions where female representation is negligible. Among them is the International Accounting Standards Board (IASB) which has adopted fair value accounting in International Financial Reporting Standards (IFRS).[31] Another influential global network is the BCBS. On most of these boards women account for 6 to 8 percent of members; the exception is the

Table 15.2 Female representation in intergovernmental, private and global regulatory institutions

Institution	*Female*	*Number of members*	*Chair*
Intergovernmental institutions			
IMF Board of Directors	4.5%	24	F[1]
World Bank Board of Directors	12%	24	M
Private sector entities			
International Accounting Standards Board (IASB)	7%	13	M
International Institute of Finance (IIF)	0%	30	M
Global committees			
International Organization of Securities Commissions (IOSCO)	30%	20	F
Basel Committee on Banking Supervision (BCBS)	8%	25	M
International Association of Insurance Supervisors (IAIS)	6%	17	M

Note:

1 Only after the executive director of the IMF, Dominique Strauss-Kahn, faced legal problems in the United States in 2011, was Christine Lagarde, the former French finance minister, elected as the new (and first female) chair of the IMF.

International Organization of Securities Commissions (IOSCO), in which women make up almost a third of the members.

Most surprising is the absence of female experts in the reform commissions, which were set up after the financial crisis with the intent of creating a new global financial architecture. Despite Albert Einstein's warning that we cannot solve problems by using the same kind of thinking we used when we created them, the financial sector relied on the same bodies that in fact were at the helm prior to the crisis. Table 15.3 demonstrates that all of the wise men and expert committee working groups set up by the G20 to prepare proposals for regulatory reforms had only male chairs. The G20 Working Group 1, which was responsible for sound regulation and improving transparency, only had 7 percent of women members (two out of a total of 29). The G20 Working Group 2 charged with reinforcing international cooperation and promoting the integrity of financial markets, had only one woman out of 27 members. Most dismal was the Working Group 3 on the reforms of the IMF in which not a single woman was a member of the total of 27. Equally disappointing was the female membership of "The Commission of Experts of the President of the UN General Assembly on Reforms of the International Monetary and Financial System," chaired by

Table 15.3 Female representation in regulatory reform initiatives

Wise men and expert committees	*Female*	*Number of members*	*Chair*
G20 Working Group 1 Enhancing Sound Regulation and Strengthening Transparency	7%	29	M
G20 Working Group 2 Reinforcing International Cooperation and Promoting Integrity in Financial Markets	4%	27	M
G20 Working Group 3 Reform of the IMF	0%	27	M
Stiglitz Commission The Commission of Experts of the President of the UN General Assembly on Reforms of the International Monetary and Financial System	9%	22	M

Joseph Stiglitz, which had two females out of a total membership of 22. This is all the more surprising since the so-called Stiglitz Commission concluded "those who are affected by the failure of regulation … should have a large voice in any regulatory structure."[32]

These empirical examples demonstrate the underrepresentation of women in top decision-making positions of key regulatory institutions, formal and informal networks. However, to reiterate the point made in the introduction, simply filling the boards with women is not sufficient. Rather, the point is to critically assess the groupthink and the abstract knowledge base that underpins finance in these club-like structures of the financial sector. The shift to largely private knowledge networks reinforces class and gender hierarchies to the detriment of the larger citizenry in developed and developing countries.

Conclusion

The conceptualization of women in economics in general, and financial economics in particular, as less rational and less capable in formal sciences has played a major role in discriminating against them in global financial networks. The excessive reliance on expert knowledge with a strong focus on quantitative modeling and the delegating of regulation to private actors has led to a situation where *regulators are captured by their knowledge base.*[33] The most striking aspect of the existing financial system is the shift of the regulatory competence to mostly private institutions, combined with the low female representation in financial market decision-making structures, a combination which, I would argue, is not coincidental because privatization has fostered regulatory capture. It is therefore not surprising that the present analysis of the origin of the crisis and the discourse and decision-making on regulatory reform is solely confined to financial insiders, mostly males, who were the culprits in the financial meltdown in the first place.

In the aftermath of the virtual meltdown of the financial system and the reflections on what caused this systemic failure, we witness both an intellectual soul searching and policy space for innovative ideas on how to create a new and more balanced global financial architecture. It is in this intellectual fluidity between old ideas and new ones not yet fully born that women experts in heterodox economics can play an important role to provide fresh ideas. There is widespread agreement that global restructuring is needed but little recognition of how to integrate alternative views of global and institutional restructuring into the contemporary discussions. What we need is a diversity of views, expertise, critical voices and a change in norms to rethink and reform the present dominance of finance in the global economy and link finance once again to the real economy. The new UN Women's agency can play an important role in this. Most importantly it should support the development of feminist economics, help feminists rethink finance, and support new models of finance that avoid the biases of the neo-liberal orthodoxy.

Essentially the question of women's inclusion as full citizens in the financial arena boils down to existing gender norms. As long as economics is constructed as a science removed from history, culture and people's lives, and as long as women are seen as less rational, less knowledgeable in formal economics and science, the field of economics in general, and finance in particular, will be unable to harness its full potential and provide benefits for all groups in society globally. A policy with redistributive consequences can only be legitimated by voters and their elected representatives.[34] Hence, any policy with redistributive impact requires input legitimacy, the incorporation of all groups in the decision-making process who are affected by financial governance. Financial regulation has been increasingly conducted beyond the political sphere, but the financial crisis has revealed significant problems of regulatory capture while excluding the interests of the public.

Women should be bold and suggest the creation of a global think tank on gender and financial governance with a mandate to work on alternative knowledge and models of financial governance that serve society best in terms of overall welfare, equity, and fairness.[35] The most important task of such a think tank is to encourage more (young) women to enter the economic fields of finance. A critical mass (of about 30 to

40 percent) of women is needed in top decision-making structures in order for alternative views to be heard by the mainstream. Creating a female expert pool is a stepping-stone to permit more women with heterodox economic skills to enter the top management positions of finance and decision-making bodies across the global economy. A further aim of the think tank would be to engage in research to ascertain whether women's behavior is different from that of men in such positions, or alternatively whether the role of women converges with that of men in such institutional settings.

A growing body of academic work in the United States and Europe suggests that there is a correlation between the number of women in company leadership and a company's profitability. While gender diversity may thus lead to greater economic efficacy, equally important is the principle of egalitarianism. The inclusion of women in top decision-making structures of the world economy is simply a prerequisite for democratic legitimacy. From this perspective, the lack of female representation in financial market regulation is problematic because the absence of women is undemocratic, while at the same time, they are dramatically affected by financial crises as we have witnessed during the Asian Crisis (1997/98), Latin American Crisis (2000/02), and the spill-over effects of these crises into other regions of the world.[36]

Notes

1 Some arguments are taken from the chapter Helene Schuberth and Brigitte Young, "The Role of Gender in Governance of the Financial Sector," in *Questioning Financial Governance from a Feminist Perspective*, ed. Brigitte Young, Isabella Bakker and Diane Elson (New York: Routledge, 2011), 249–92.

2 Susan Strange, "The Study of International Political Economy," in *Theorien Internationaler Politik*, ed. Ursula Lehmkuhl (München/Wien: R. Oldenburg Verlag, 1994), 310–30.

3 Strange, "The Study of International Political Economy," 323.

4 Brigitte Young, Isabella Bakker and Diane Elson, "Introduction," in *Questioning Financial Governance from a Feminist Perspective*, ed. Brigitte Young, Isabella Bakker, and Diane Elson (New York: Routledge, 2011).

5 Philip Cerny, "Globalisation and Statehood," in *Issues in 21st Century World Politics*, ed. Mark Besson and Nick Bisley (Houndsmills: Palgrave Macmillan, 2010), 17–32.

6 Schuberth and Young, "The Role of Gender in Governance of the Financial Sector."

7 Strange, "The Study of International Political Economy," 310–30.

8 Diane Stone, "Knowledge and Policy Networks in Global Governance," in *Gender Knowledge and Knowledge Networks in International Political Economy*, ed. Brigitte Young and Christoph Scherrer (Baden-Baden: Nomos, 2010), 36–54.

9 Peter Mooslechner, Helene Schuberth and Beat Weber, *The Political Economy of Financial Market Regulation. The Dynamics of Inclusion and Exclusion* (Cheltenham/Northampton: Edward Elgar, 2006), xii–xxviii.
10 Strange, "The Study of International Political Economy," 318.
11 Eugene Fama, "Efficient Capital Markets: A Review of Theory and Empirical Work," *Journal of Finance* 25, no. 2 (1970): 383–417.
12 Robert Cox, "Social Forces, States, and World Orders: Beyond International Relations Theory (1981)," in Robert Cox with Timothy J. Sinclair, *Approaches to World Order* (Cambridge: Cambridge University, 1996), 85–123.
13 Isabella Bakker, *The Strategic Silence: Gender and Economic Policy* (London: Zed Books, 1994).
14 Brigitte Young and Christoph Scherrer, "Introduction: Gender Knowledge and Knowledge Networks," in *Gender Knowledge and Knowledge Networks in International Political Economy*, ed. Brigitte Young and Christoph Scherrer (Baden-Baden: Nomos, 2010), 9–17.
15 Diane Elson and Nilufur Cagatay, "The Social Content of Macroeconomic Policies," *World Development* 28, no. 7 (2000): 1347–64.
16 Brigitte Young, "The 'Mistress' and the 'Maid' in the Globalized Economy," in *Socialist Register 2001*, ed. Leo Panitch and Colin Leys (Pontypool: Merlin Press, 2000), 315–28.
17 Schuberth and Young, "The Role of Gender in Governance of the Financial Sector"; and Brigitte Young, "The Gendered Dimension of Money, Finance and the Subprime Crisis," in *Gender and Economics. Feminstische Kritik der politischen Ökonomie*, ed. Christine Bauhardt and Caglar Gülay (Wiesbaden: VS Verlag, 2010), 203–18.
18 Paul Langley, "When Credit Becomes Debt: Foreclosures and Forbearance in Subprime Crisis—The Economic, Politics and Ethics of Response" (Conference Paper, Warwick University, 18–19 September 2008); also Stephen Gill and Adrienne Roberts, "Macroeconomic Governance, Gendered Inequality, and Global Crisis," in *Questioning Financial Governance from a Feminist Perspective*, ed. Brigitte Young, Isabella Bakker, and Diane Elson (New York: Routledge, 2011), 155–72.
19 Eleni Tsingou, "The Study of Global Finance in International Political Economy: Introducing Gender—a Research Agenda," in *Gender Knowledge and Knowledge Networks in International Political Economy*, ed. Brigitte Young and Christoph Scherrer (Baden-Baden: Nomos, 2010), 150–59.
20 Stone, "Knowledge and Policy Networks in Global Governance," 36.
21 The idea of "groupthink" is not new. David Halberstam in *The Best and the Brightest* (Greenwich, CT: Fawcett Crest Books, 1973) describes the phenomenon of the Kennedy intellectuals, who were praised as the best and the brightest men of a generation and yet they as a group were also the architects of one of the worst tragedies of the United States to believe that bombing was to achieve victory in Vietnam.
22 Gillian Tett, *Fool's Gold. How Unrestrained Greed Corrupted a Dream, Shattered Global Markets and Unleashed a Catastrophe* (London: Little, Brown Book Group, 2009).
23 Paul Krugman, "How Did Economists Get It So Wrong?" *The New York Times*, 6 September 2009, www.nytimes.com/2009/09/06/magazine/06Economic-t.htm?

24 Peter S. Goodman, "Taking Hard New Look at a Greenspan Legacy," *The New York Times*, 8 October 2008, http://www.nytimes.com/2008/10/09/business/economy/09greenspan.html?pagewanted=all.
25 Martin Hellwig, "Systemic Risk in the Financial Sector: An Analysis of the Subprime-Mortgage Financial Crisis," Working Paper Series of the Max Planck Institute for Research on Collective Goods, no. 43 (2008), Cologne: Max Planck Institute.
26 Timothy Sinclair, *The New Masters of Capital. American Bond Rating Agencies and the Politics of Creditworthiness.* Cornell Studies in Political Economy (Ithaca, NY: Cornell University Press, 2008).
27 Vanessa Redak, "Risks, Rating and Regulation: Toward a Reorganization of Credit via Basel II?" in *The Political Economy of Financial Market Regulation. The Dynamics of Inclusion and Exclusion*, ed. Peter Mooslechner, Helene Schuberth, and Beat Weber (Cheltenham/Northampton: Edward Elgar, 2006): 191–213.
28 William Vleck, "Surveillance to Combat Terrorist Financing in Europe: Whose Liberty, Whose Security?" in *European Security* 16, no. 1 (2007): 99–119, http://www.ssrnetwork.net/document_library/printable_document_detail.php?id=3765.
29 Stijn Verhelst, *Renewed Financial Supervision in Europe—Final or Transitory?,* Egmont Papers 44 (Brussels: The Royal Institute for International Relations, March 2011).
30 Tonia Warneke, "A Gender-aware Approach to International Finance," in *Ethics and the Market Insights from Social Economics*, ed. Betsy Jane Clary, Wilfred Dolfsma and Deborah M. Figard (New York: Routledge, 2006), 175–90.
31 Juan A. Solé, Alicia Novoa, and Jodi Scarlata, "Procyclicality and Fair Value Accounting," IMF Working Paper no. 09/39 (2009), http://papers.ssrn.com/sol3/papers.cfm?abstract_id=1366168.
32 Joseph Stiglitz, "Principles for a New Financial Architecture. The Commission of Experts of the President of the UN General Assembly on Reforms of the International Monetary and Financial System," www.un-ngls.org/docs/ga/cfr/newfinancialarchitecture.pdf.
33 Nicholas Dorn, "Ponzi Finance, Regulatory Capture and the Credit Crunch," 19 March 2009, http://ssrn.com/abstract=1365250.
34 Giandomenico Majone, "Independent Agencies and the Delegation Problem: Theoretical and Normative Dimensions," in *Political Institutions and Public Policy. Perspectives on European Decision Making*, ed. Bernand Steunenberg and Frans van Vught (Dordrecht: Kluwer Academic Publishers), 139–56.
35 Brigitte Young and Helene Schubert, "The Global Financial Meltdown and the Impact of Financial Governance on Gender," *GARNET Policy Brief* no. 10 (January 2010).
36 Joseph Limm, "The Effects of the East Asian Crisis on the Employment of Women and Men: The Philippine Case," *World Development* 28, no. 7 (2000): 1285–1306; also Ajit Singh and Ann Zammit, "International Capital Flows: Identifying the Gender Dimensions," *World Development* 28, no. 7 (2000): 1249–68; and Brigitte Young, "Financial Crisis and Social Reproduction: Asia, Argentina and Brazil," in *Power, Production and Social Reproduction*, ed. Isabella Bakker and Stephen Gill (Houndsmills: Palgrave Macmillan, 2003), 103–23.

16 Conclusion

Advancing feminist strategies in international governance

Gülay Caglar, Elisabeth Prügl and Susanne Zwingel

- **Findings**
- **Implications for future research and theory building**
- **Policy implications**

The contributions in this book examine a long period of feminist interventions in international governance based on a variety of strategies, goals, and actor constellations. This conclusion is an attempt to summarize our contributors' findings. Furthermore, we will try to tease out research gaps, discuss the theoretical implications of our findings and outline paths for future research. Finally, we will elaborate on the policy implications we see emerge for UN Women from the diverse range of academic literature represented in this book.

Findings

What answers do contributors to this book provide to the questions initially raised: How has global recognition of gender equality become possible, how far does this recognition reach, and how profound are the transformations it has triggered around the world?

Remarkably and with only one exception in the field of financial governance, all chapters in this volume describe cases and processes where women have managed to get a foot in the door of international governance institutions and have been able to enshrine gender equality into legislative and policy practices at the international level. Feminist ideas have made some kind of an impact; however, the degree to which this happens in a meaningful way depends on many factors. One important factor is the in- or exclusiveness of the policy field. Of the three fields for which our authors analyze the reach of feminist interventions, the human rights regime has been, normatively speaking, the most inclusive. This is not surprising because one of its goals is non-discrimination on all grounds and dignified living conditions for all, including women. In contrast, security and economic governance are more exclusive: Both fields are guided by ideational traditions that ignore gender as a relevant dimension when working for desired goals such as security, peace, economic prosperity or poverty eradication. Thus, we may be more ready to expect an institutional culture that sidelines feminist claims in these fields, and the contributions of von Braunmühl, Razavi, Bedford, Caglar, Young, and Williams confirm this expectation.

The field of finance is particularly resistant to including women and to feminist interventions. Young's analysis of finance and financial governance institutions shows a striking absence of women and pervasive gender bias in the orthodox models applied. Her analysis suggests that crisis-producing financial policies were the result of an exclusive construction of "expertise" based on particular (male, Western, wealthy) perspectives and interests. Accordingly, she calls for both numerical representation of women and the substantial presence of feminist and otherwise inclusive ideas as a way of democratizing and ultimately improving financial governance. Her proposal is reminiscent of the claims of nineteenth- and early twentieth-century suffragettes: They asked for representation in a political setting that justified their exclusion based on their innate inability to "rationally" decide over public matters. Young's analysis reminds us that the exclusion of women is not just a problem of previous centuries, but that the patterns of legitimizing such exclusion have changed. Present-day exclusion is rarely based on blatant essentialism, but on the construction of superiority of a certain type of expertise that elite men are most likely to acquire.

The feminist record outside finance is more impressive, leading to our first question, i.e., how has the recognition of gender equality become possible? For the most part, strategies of lobbying and agenda setting that dominated literature in the 1980s and 1990s—documenting the activism and achievements around the UN Women's conferences—are no longer the central focus of analysis for the contributions in this volume. Rather, our contributors refer to and build on the legacy of feminist activism from earlier decades: the existence of equality legislation, policy documents such as the Beijing Platform for Action (BPA), and the more recent Security Council (SC) resolutions on gender. They also accept that most international institutions are committed to gender mainstreaming in some form or other and that there are gender experts in institutions who drive agendas. The only area where our authors describe strategies for getting gender on the agenda is international economic governance. Williams documents the difficulty of feminist strategizing in the field of international trade policy and Caglar illustrates the power of feminist knowledge networking to move agendas in the context of macro-economic policy. The difficulty of integrating gender into macro-economics stands in stark contrast to the embracing of gender issues in development economics, in particular through the World Bank. Both Razavi and Bedford remind us that this large-scale integration of gender became possible only after the modification of the Washington Consensus and the rediscovery of social policy in a new form. In other words, agenda setting required the right institutional and discursive opportunity structure.

Thus, agenda setting is recognized as a demanding first step. The focus of most of our contributions is, however, on analyzing the continuous strategic work that is required to anchor feminist ideas in international organizations and to maintain relevance of those ideas in organizational practice. Several authors elaborate on the complexities and challenges of getting feminist ideas heard and keeping them relevant within the institution. Hannan elaborates on the cumbersome work of integrating gender equality claims into United Nations (UN) agendas. It requires framing of issues in ways acceptable within a male-dominated organization while insisting on core feminist goals; coalition building with powerful "willing actors" within the organization and continuous persuasion on several levels; awareness of the limits and opportunities an intergovernmental bureaucracy has to offer; and openness to input from women's organizations to maintain a fresh outlook on things. Caglar explores the ways in which networks of feminist economists have initiated complex learning processes as to the relevance of gender-sensitivity in economic governance institutions. Similarly to Hannan, she identifies the importance of careful framing, that is, the presentation of new issues in ways acceptable for the institution yet without sacrificing their transformative potential.

Moreover, Hannan describes the need to defend far-reaching concepts represented in the Convention on the Elimination of All Forms of Discrimination against Women (CEDAW) and the BPA in conservative times. Such times did not allow pushing controversial issues forward, but required femocrats' full energy to defend what was already agreed upon. Zwingel finds for the CEDAW convention that its adoption—without doubt a case of successful agenda setting—in itself did not create intergovernmental recognition of the instrument. Rather, such recognition has required ongoing efforts of awareness raising and lobbying undertaken by committee members, UN bureaucrats, supportive government delegations, and a limited but influential number of nongovernmental organizations (NGOs).

The strategy of gender mainstreaming has become the most widely used instrument of institutionalizing gender awareness in international organizations. Critical reflections of its rather superficial application indicate, however, that continuous feminist awareness raising is necessary to make gender mainstreaming as transformative as it is meant to be (True and Parisi). In this context, the influence of gender expertise is an important attempt to anchor a substantial understanding of gender within organizational practice. Gender expertise develops gendered solutions to various ends pursued in different issue areas drawing on the power of essentialist reasoning, the discourse of equality and human rights, and statistical and legal tools. Prügl reminds us that such expertise, while giving authority to feminist ideas, also has the potential of de-radicalization, in part because feminist ideas are passed on from the top as "the right medicine" rather than owned and autonomously developed by women's rights activists.

Together, these findings suggest that creating and maintaining recognition for gender equality claims within institutions of international governance is hard and never ending work. But where the recognition has occurred, how far has it gone? This was our second question and the one ultimately most central to the contributions in this book. Including feminist ideas into governmental institutions has produced contradictions and unintended consequences with which both practitioners and academics grapple.

The most notable tension is the repeatedly identified need for collaboration, bridge-building, acceptable framing of issues on the one hand and the need to protect core feminist ideas from being co-opted on the other. Authors often describe strategies based on networking with other actors and persuading them of the validity of their goals. Where the goal is getting feminist ideas onto the agenda of international organizations or pushing them further, persuasion is crucial, and knowing how to play according to intergovernmental as well as field-specific rules is a strategic asset. Networking also happens between ideational coalition partners, e.g., NGOs and femocrats or among a group of NGOs. This kind of activism aims at strengthening a position and often also at ensuring an autonomous voice that is less likely to be co-opted (e.g., Williams). While coalition building is an instrument to increase the base for a claim, it always contains the need to find a common denominator and, as a result, to compromise. While some contributors describe this dynamic as an integral part of policy making (e.g., Hannan), others express frustration about the loss of substance implied: Charlesworth cannot find sufficient advocacy for women's issues within the human rights framework, partly because integrating gender has mostly meant to single out women as a partial group vis-à-vis the general male comparator. Von Braunmühl sees the efforts of feminist coalitions in the security sector seep away into an integration of women-as-victims, but not participants and agents. Jenichen finds this same construction in representatives of international organizations deployed in post-war conflict zones: while they should be the ones integrating gender-awareness in the post-war reconstruction process, they only solidify "traditional" gender norms they expect local communities to follow. Razavi shows that when gender has been integrated into new development thinking without questioning neo-liberal basics, it has generated the "social investment perspective," which has put investment in children at the center, subsuming women's needs under those of their children. Bedford adds that this approach has tended to reformulate the feminist debate about a crisis in social reproduction to a crisis of the family, generating problematic policies that (re)privatize care while reconstructing traditional, heterosexual couple relations. Williams's contribution on nongovernmental efforts to influence the World Trade Organization (WTO) agenda is perhaps most decidedly critical toward the dangers of co-optation. The strategic work she describes and reflects on is very much based on broad awareness raising and coalition building outside of the WTO in order to become a voice that is heard inside; yet she formulates opposition toward the effort to mainstream gender into the institutional structure of the WTO as the WTO is driven by an ideology that is too far away from the feminist goal of creating an emancipatory and gender-sensitive trade regime. Thus, a general conclusion to be drawn from this collection in regard to the tension between cooperation and co-optation is that without critical feminist assessments—ideally coming from autonomous actors—any attempt to meaningfully engender international governance institutions is likely to remain superficial and probably not worth the effort.

Finally, the contributions in this volume suggest that our last question on the degree of worldwide repercussions of the global recognition of gender equality has been least systematically addressed by academic research. This is not surprising as such repercussion is not easy to conceptualize or measure. However, a range of contributions do explore the connection between impulses of international governance and other-than-global contexts. Read together, they show that recognition of gender equality takes on manifold meanings once the scope of analysis is expanded from intergovernmental to domestic contexts as a variety of local interests and framings come into play. In this vein, Levitt and colleagues show how two nongovernmental actors in Peru, one embedded in liberation theology, the other in feminism, differently integrate global ideas about women's rights into their work: The faith-based organization SEA aims at ensuring the rights of women (as "children of God") within the scope of Catholic thought in organizing collective initiatives such as communal kitchens whereas the feminist organization DEMUS puts emphasis on the reproductive rights of women (as autonomous subjects). Sabat stresses the obstacles for implementing gender norms in Lebanon. She identifies a number of domestically dominant convictions, i.e., in regard to family honor and family cohesion that challenge the goal of gender equality and, as a consequence, require the adaptation of some elements of gender norms to the local context. In their analysis of the Swedish, British and German national action plans to implement SC Resolution 1325, Schneiker and Joachim also find a variety of interpretations of the UN gender and security framework. Sweden and the United Kingdom have a more far-reaching understanding of the framework than Germany, which is not only reflected in the substantial measures laid out in the plan, but also in the participatory process of producing the document. The authors conclude that the single most important factor for Sweden's and Great Britain's substantial translation of the resolution into domestic practice was a domestic actor constellation consisting of both feminism-savvy governmental actors and independent feminist experts and activists. Jenichen's analysis of successful feminist mobilization in post-conflict Bosnia Herzegovina yields an interesting and maybe unexpected result: Domestic women's rights activists may be much more aware of global gender norms than international civil servants present to facilitate a transition to sustainable peace. She demonstrates how a rather weak group of domestic and international feminist advocates came together to educate domestic and international decision makers and ultimately got support for some core feminist claims. In sum, as domestic actor constellations vary widely so does the way gender equality norms are interpreted, recognized, and realized.

Implications for future research and theory building

One of the biggest gaps in current research pertains to exploring the impacts of feminist strategies in international governance. In particular, the impacts of gender mainstreaming are difficult to ascertain. As True and Parisi show, a lot is known about the conditions that need to be in place for gender mainstreaming to be implemented, about the way in which gender mainstreaming produces unintended outcomes and co-optations, and about the way in which organizations undermine the transformative intent of the strategy. But relatively little is known about what this actually means for women and men in different societies. What are the effects of checklists, gender analyses, gender budgets and policies on these women and men? Another way to ask this question would be: How do state policies generate changes in society?

One way to begin to address this question is the development of statistical indicators that has garnered new attention in academia and in policy circles. Among the newer indicators are the annual Global Gender Gap Reports of the World Economic Forum (WEF), initiated in 2006, and the United Nations Development Programme's (UNDP) Gender Inequality Index, presented for the first time in the 2010 Human Development Report. Both account for the conceptual deficiencies of the UNDP's earlier Gender-related Development Index (GDI) in different ways.[1] The WEF's gender gap index is particularly innovative in that it focuses on gender inequality independently of a country's human development level. It specifies an ideal target of gender equality and measures the degree to which countries have approximated that target, that is the degree to which they have closed their gender gaps. Overall, gender indices provide evidence of developments over time and in this way a tool to gauge the potential impact of gender equality policies.

However, there are difficulties. To begin with, it is difficult to causally link domestic developments toward more (or less) gender equality to the impacts of international gender norms, policies and the manifold strategic interventions analyzed in this volume. Changes in gender relations may result from multiple sources, which may or may not be connected to international politics. Furthermore, the ranking of countries changes depending on the method of measurement. This signals that the rankings themselves may contribute to constructing a reality of gender equality and inequality. They end up defining equality, generate standards by which countries are held accountable, implicitly foreclose alternative understandings of what gender equality may mean, and abstractly stabilize a concept that is a matter of considerable and highly politicized debate in many local contexts. Despite these difficulties, there is a strong need for creative quantitative analyses that begin to collect data on the implementation of international gender equality norms, conceptualize causal pathways and begin to establish correlations between international influence, governmental policies in support of gender equality, and impacts of these on gender relations generally and the status of women specifically.

A difficulty in evaluating the impacts of international gender norms lies in the fact that its goals are unclear—True and Parisi make this particularly explicit in their assessment of gender mainstreaming. Implicitly the UNDP and WEF indices come to define the goals of feminist struggle. Yet, this solution can only be preliminary as the concept of gender equality is essentially contested. It is a matter of debate not only among international agencies who co-opt the issue for their purposes, but also among feminists who disagree on whether feminists should strive for sameness or a celebration of difference, and who debate whether, given the fact that truths invariably are situated, a universal standard of gender equality should be defined. In light of these debates, studies of the impact of international gender equality policies cannot stop with quantitative indicators defining the dependent variable.

Although True and Parisi also are critical of a tendency to focus on the processes of gender mainstreaming rather than the goals, the uncertainty about such goals makes a study of processes imperative. We believe that research on the impacts of women's human rights law can be instructive here for various reasons. First, this research has probed impacts from the bottom up by focusing on the interactions of local politics with the normative resources provided by CEDAW and other human rights instruments. In doing so, it has linked up with research in comparative politics to explore how national institutions (including social movements and women's policy machineries) are able to serve as conduits for international norms. In part, it has drawn on anthropological approaches and ethnographic methods to explore the impact of norms from the bottom up (see for example, Levitt *et al.*, Sabat, Jenichen in this book). Applied to gender mainstreaming, this approach would provide localized answers, showing intended and unintended effects and illustrating complex mechanisms unleashed through the deployment of feminist strategies. Second, studies on the local impacts of international women's rights law challenge the assumption that a negotiated international norm has a strategically intended effect. Instead they make room for domestic politics and for the translations and interpretations that invariably happen when norms travel (Zwingel). The meaning of a norm derives not from an imagined codified intention but from the process of norm translation. In these approaches then, international gender politics becomes a terrain of meaning construction, international policies fluid propositions that need to gain reception and pass through societal filters before they become part of the vernacular. Such studies presume not a world of rational public administration but a world of politics, buffeted by divergent interpretations, interests, resistances, dealings, and negotiations. The meaning of gender equality is defined in this world of politics. Indices and gender gap reports may suggest what gender equality means, but they are just one political force among many. Furthermore, in this world the impact of international politics becomes visible in the engagements of local actors with international norms and discourses. Impact is not a matter of measuring changes in women's labor force participation or political participation, but a matter of local actors adapting and embracing propositions put forward by gender experts.

This conceptualization of international gender politics links up to the global governmentality approach. Both emphasize the discursive character of such gender politics and recognize its involvement in the government of gender on a global scale (Bedford, Caglar, Prügl). From this perspective it becomes difficult to distinguish true feminist goals from the discourse deployed in gender mainstreaming or in arguing legal cases, or in setting agendas and implementing them. In other words, the global governmentality approach precludes holding actors accountable to an imagined true standard of achieving gender equality. But the approach opens up other interesting venues for critical inquiry. First, it recognizes that the goals of gender politics are derived from the construction of the problem thereby illustrating that both could be formulated differently. In this way, it provides another access to the notion of policy framing and the issue of co-optation. Second, discourses can usefully be juxtaposed and thus provide an analytical wedge to clarify varied political effects. Thus, gendered discourses need to be held up not to an abstract feminist goal but can be measured against each other. Third, a focus on discourses makes visible a layer of politics not captured by interests, yet equally suggestive, i.e., a politics of purpose encoded in discursive commitments. Thinking of international feminist politics in this way suggests a different way of conceptualizing strategies and their impacts. If strategies are always already inherent in discourses, feminist politics may mean making such strategies visible together with the effects they produce in order to gauge impacts. Bedford illustrates how to do so when she uncovers the way in which the discourse on the family that feminist framers of PROFAM had put forward for political purposes separated itself from their intentions to produce relatively conservative realities.

Policy implications

This book was inspired by the creation of UN Women, which motivated us to rethink what academics know about international gender politics and to develop the implications of this knowledge for the new agency. Our contributors seem to come together around four recommendations. The first pertains to UN Women as a think tank, the second to ensuring civil society participation, the third to strengthening gender mainstreaming, and the fourth to linking the women's rights framework to other realms of gender policy work. We will discuss each of these recommendations in turn.

The first and most important emphasis of the authors in this volume is their appeal to UN Women to serve as a think tank on international gender issues. The theme takes various forms. On the one hand, those who study issue areas that have resisted the incorporation of gender considerations are calling on the agency to help develop new concepts, analytical instruments, and framings. The need is perhaps most dire in the field of finance, where the crises of 2008 have spawned a recognition of the extreme masculinity of the field, and feminist analysts are now beginning to look at the field through gender lenses (Young). But there also is considerable work left in the fields of macroeconomics and trade, where feminist models have been developed (in part with United Nations Development Fund for Women support), but their impact remains marginal. Williams emphasizes the importance of analytical models for the advocacy work of International Gender and Trade Network. Caglar shows that changes in the field of international economic governance have been incremental but not paradigmatic due to the inability of international organizations to learn from feminist economists and to apply their analytical models. Yet as long as international organizations are unable to learn, they will continue to regard gender politics in economic governance as a social issue separated from economic policy. In a similar fashion, Jenichen pleads for a serious investment in gender awareness in the Department of Peacekeeping Operations. In all these fields, UN Women needs to help develop new analytical models and actively advocate for their uptake in UN departments and specialized agencies.

But UN Women is called on to serve as a think tank in other, more established issue areas as well. Hilary Charlesworth, for example, emphasizes the need to go beyond the discourse of non-discrimination by developing new concepts such as the notion of the undervaluation of women's work (rather than unequal pay) and the notion of policy neglect (for example in the monetary field). These kinds of framings provide argumentative punch in directions that are necessary to address the specific situations of women. They also provide a way to document the insufficiency of existing legal interventions. Charlesworth in this way links up to Bedford's recommendation on the necessity to remain curious about all levels of governance, to continue to investigate the gendered objectives that may hide behind a seemingly neutral policy approach (such as macroeconomics), and to be critical of what she calls "gender sharing." Another way to formulate the recommendation of these authors more broadly might be to ask UN Women to become a resource for feminist reflexivity, for remaining sensitive to the power that it and other feminist experts exercise in their work, to foster analytical work that makes this power visible, and to provide a space for reflection on the operations of power in feminist governmentality. In doing so, UN Women can be at the forefront of showcasing how to apply reflexivity in governance and provide a model for more democratic policy-making for international institutions in general.

The second recommendation of our authors is for UN Women to work with civil society and include such work in all areas of its operation. True and Parisi claim that women's movements have been virtually excluded from gender mainstreaming. It is left to bureaucrats to interpret and apply gender equality policy without any input from civil society. Gender mainstreaming is weakened by keeping out feminist activists and loses legitimacy. It also is less effective than it could be. Razavi argues that working with civil society is crucial to a strategy of engendering economic governance. For her, civil society is not limited to international NGOs, but needs to include social forces from all levels of international governance, including national and sub-national levels. Zwingel, Levitt *et al.*, Sabat, and Jenichen all highlight the important role that NGOs are playing in advancing gender equality rights. The promise of including civil society in UN Women's work is thus increasing legitimacy and effectiveness in the advancement of gender equality goals.

Despite their far-ranging critique of gender mainstreaming, our authors come around to the suggestion that the strategy needs to be strengthened rather than abandoned and call on UN Women to help do so. Hannan emphasizes that gender mainstreaming has never been fully implemented in the UN and that there continue to be significant shortages of capacity. She calls for an acceleration of implementation. True and Parisi are highly critical of the way in which gender mainstreaming has been implemented, but support the commitment of Michelle Bachelet for more of the strategy. Their recommendation is that UN Women should make sure that the transformative intent of gender mainstreaming be realized. More carefully, von Braunmühl argues for a realistic rethinking of what gender mainstreaming can accomplish. In a hopeful note on the potential for gender mainstreaming, Williams recognizes the important role that gender machineries could play in inserting gender considerations into trade negotiations. The challenge for UN Women will be to proliferate a strategy that has proven highly problematic while at the same time holding governments and international organizations accountable on the impacts of the practice.

The fact that UN Women combines a policy-making, operational and monitoring role provides it an opportunity for demanding more accountability from governments, the UN, and other international agencies. In doing so, we believe that the agency should draw strongly on the women's rights framework and on the work of CEDAW. As the contributions by Levitt *et al.* and Sabat show, this framework already does important work in addressing women's rights and delegitimizing gender-based violence in a variety of contexts. However, it also holds considerable potential for advancing economic rights and social justice. As Levitt *et al.* argue, activists already are drawing on the women's rights framework to argue for social justice. In this way, human rights do not operate only as legal instruments but as a political ideology and social movement. Working with CEDAW and human rights instruments thus provides a powerful political tool for UN Women, which it will need to supplement with linkages and allies across the UN and in academia. Zwingel pleads for UN Women to prioritize connecting disconnected contexts, including a focus on intra-organizational coherence within the UN.

In brief, UN Women can play a crucial role for both creating strong feminist voices within the UN and strengthening gender norms throughout the world. Proper funding is an important prerequisite for attaining these objectives. The funding commitments of governments show that it is still a challenge to resource international gender politics adequately.[2] It is important, though, that UN Women has the structural as well as the financial ability to fulfill its role as a pivot in international gender politics; that is to support innovative projects, to connect to the outside world of feminist and women's movements and academia for uncompromised input and, ultimately, to serve as a critical think tank.

Getting feminist concerns into international governance is, as the contributions in this volume show, an ongoing process which can be characterized as a two steps forward, one step back project. Feminist actors throughout the UN have persevered while organizational logics have sidelined, sidetracked and resisted feminist intentions. Thus international gender politics and feminist strategies continuously need to be evaluated from a critical point of view. It is important to strengthen the dialogue between institutional actors, working on integrating gender into institutions, and feminist activists in order to keep an empowerment-oriented, critical feminist discourse alive and thereby enable transformative gender politics. This book wants to contribute to such a dialogue.

Notes

1 For further elaborations see Ricardo Hausmann, Laura D. Tyson, and Saadia Zahidi, *Global Gender Gap Report 2007* (Geneva: World Economic Forum, 2007), 3, and http://hdr.undp.org/en/statistics/gii/.

2 Jane Martinson, "UN Women Justice Report: Gender Equality Programs Need Funding," *The Guardian*. 6 July 2011, http://www.guardian.co.uk/global-development/2011/jul/06/un-women-gender-equality-underfunded.

Select bibliography

Amrita Basu, ed., *The Challenge of Local Feminisms. Women's Movements in Global Perspective* (Boulder, CO: Westview Press, 1995), and *Women's Movements in a Global Era. The Power of Local Feminisms* (Boulder, CO: Westview Press, 2010).

Both volumes analyze diverse local struggles and transnational politics in reaction to intergovernmental impulses and broader global (e.g. economic) challenges.

Kate Bedford, *Developing Partnerships: Gender, Sexuality, and the Reformed World Bank* (Minneapolis, MN: University of Minnesota Press, 2009).

Through a case study of Word Bank policies, the book shows how global economic and gender equality policies participate in enforcing expectations about sexuality, gender and family relations.

Gülay Çaglar, *Engendering der Makroökonomie und Handelspolitik. Potenziale transnationaler Wissensnetzwerke* (Wiesbaden: VS Verlag für Sozialwissenschaften, 2009).

Tracing the knowledge networks of feminist economists seeking to influence macroeconomic models in various international forums, the book illustrates various logics underlying these feminist interventions.

Hilary Charlesworth and Christine Chinkin, *The Boundaries of International Law: A Feminist Analysis* (Manchester: Manchester University Press, 2000).

This book sheds light on the ways in which international law has failed to address the marginalization of women around the world. It aims at redrawing the boundaries of international law in order to make it more responsive to gender-related injustices.

Diane Elson and Nilüfer Çagatay, "The Social Content of Macroeconomic Policies," *World Development* 28, no. 7 (2000): 1347–64.

The article criticizes macroeconomic policymaking and illustrates how to mainstream gender-equitable social issues into macroeconomic policies. It argues that a participatory approach to economic policymaking is an important prerequisite for formulating alternatives.

Anne-Marie Goetz, ed., *Getting Institutions Right for Women in Development* (London: Zed Books, 1998).

This book offers a critical analysis of the range of gender policies employed by development organizations. It argues that such policies have helped improve the immediate material situation of women, but by and large have contributed to reproducing the ideological and material conditions for women's subordination in the family and the economy.

Mary E. Hawkesworth, *Globalization and Feminist Activism* (Lanham, MD: Rowman & Littlefield Publishers, 2006).

This book shows that globalization is a gendered phenomenon and illustrates how feminist activists have worked for global transformation. Analyzing innovative feminist strategies and tactics to produce global change, the book traces the structural forces that permeate and constrain transnational feminist activism.

Rosemary Hunter, ed., *Rethinking Equality Projects in Law: Feminist Challenges* (Oxford: Hart Publishing, 2008).

This collection offers a re-examination of the widely applied concept of equality as a tool for improving the status of women. The contributions discuss both competing conceptions of equality and alternatives to the concept as more suitable ways to achieve feminist goals.

Rounaq Jahan, *The Elusive Agenda: Mainstreaming Women in Development* (London: Zed Books, 1995).

This book reviews policies and measures to promote women's advancement in a variety of development contexts and multilateral donor agencies. It argues that the objective of transforming social and gender relations and creating a more just and equitable world is far from being achieved.

Devaki Jain, *Women, Development and the United Nations: A Sixty Year Quest for Justice* (Indianapolis, IN: Indiana University Press, 2005).

The author reviews the evolution of the UN's programs aimed at benefiting women and the impact of women's ideas about rights, equality, and social justice on UN thinking and practice regarding development.

Jutta Joachim, *Agenda Setting, the UN, and NGOs: Gender Violence and Reproductive Rights* (Washington, DC: Georgetown University Press, 2007).

This book shows the institutional and normative framework in which transnational women's networks are embedded and which forms of "strategic framing" and "discursive politics" they develop to legitimize certain gender equality claims at international and local levels.

Nüket Kardam, "The Emerging Global Gender Equality Regime from Neoliberal and Constructivist Perspectives in International Relations," *International Feminist Journal of Politics* 6, no.1 (2004): 85–109.

This article identifies and describes the emergence of a global gender equality regime, a process in which global women's networks and multilateral and bilateral development organizations have been instrumental. It also points to the continuing interpretation and contestation of the meaning of gender norms within this regime.

Dyan E. Mazurana, Angela Raven-Roberts and Jane Parpart, ed., *Gender, Conflict, and Peacekeeping* (Lanham, MD: Rowman & Littlefield, 2005).

This volume offers an analysis of the key role of gender in post-conflict peacekeeping efforts and explores how gender has become a central factor in shaping current thinking about the causes and consequences of armed conflict. It draws on expertise ranging from the highest levels of international policymaking to the daily struggle of implementing peacekeeping operations.

Sally E. Merry, *Human Rights and Gender Violence: Translating International Law into Local Justice* (Chicago, IL: University of Chicago Press, 2006).

This book makes the argument that international human rights law needs to be framed in local terms to be accepted and effective. It uses the phenomenon of violence against women to show the tensions that exist between global legal principles and local practices.

Mary K. Meyer and Elisabeth Prügl, ed., *Gender Politics in Global Governance* (Lanham, MD: Rowman & Littlefield Publishers, 1999).

The edited volume illustrates how feminists have been able to create spaces and to incorporate gender concerns within international organizations and how women's movements have influenced international policy processes and norms.

Valentine M. Moghadam, *Globalizing Women: Transnational Feminist Networks* (Baltimore, MD: Johns Hopkins University Press, 2005).

Introducing various transnational feminist networks—their foundations, objectives and programs—the book examines how such networks are responding to and resisting growing inequalities, the exploitation of female labor, and patriarchal fundamentalisms.

Donna Pankhurst, ed., *Gendered Peace: Women's Struggles for Post-War Justice and Reconciliation* (New York: Routledge, 2007).

This book explores struggles for gender awareness in post-conflict settings. It highlights the surprising success in the development of international legal advances for women and contrasts this with the actual experience of women in a variety of cases from Sierra Leone to the Balkans.

Elisabeth Prügl, *The Global Construction of Gender. Home-Based Work in the Political Economy of the 20th Century* (New York: Columbia University Press, 1999).

This book explores the debates and rhetoric surrounding home-based workers that have taken place in global movements and multilateral organizations since the early 1900s in order to trace changing conceptions of gender over the course of this century.

Shirin M. Rai and Georgina Waylen, ed., *Global Governance. Feminist Perspectives* (Basingstoke, New York: Palgrave MacMillan, 2008).

This book provides a comprehensive analysis of global governance from a gendered perspective. It furthers the emerging feminist theorizing on global governance and provides a theoretically informed and empirically based analysis of both institutions and transformative practices.

Judith Squires, *The New Politics of Gender Equality* (Basingstoke, New York: Palgrave MacMillan, 2007).

This book analyzes the potential of gender equality policies such as women's policy agencies, gender quotas and gender mainstreaming. It argues that they need to be framed by considerations of democratic justice rather than technocratic utility and complex diversity rather than sexual difference.

Mieke Verloo, ed., *Multiple Meanings of Gender Equality. A Critical Frame Analysis of Gender Policies in Europe* (Budapest: Central European University Press, 2007).

This book aims to map the diversity of meanings of gender equality across Europe. It uses Critical Frame Analysis to explore the different interpretations of gender equality in a variety of geographical locations and in relation to diverse issues including family policies, domestic violence, prostitution and migration.

Brigitte Young, Isabella Bakker and Diane Elson, ed., *Questioning Financial Governance from a Feminist Perspective.* IAFFE Advances in Feminist Economics (London, New York: Routledge).

This volume sheds light on the gender dimensions of macro-economic and financial governance.

Susanne Zwingel, "How Do International Norms Travel? Theorizing International Women's Rights in Transnational Perspective," *International Studies Quarterly*, 56, no. 1 (2012): 1–15.

This article develops a transnational perspective on norm translation as a multidirectional process of appropriation and contestation. It uses this approach to theorize repercussions of the Convention on the Elimination of All Forms of Discrimination Against Women (CEDAW) as the centerpiece of the international women's rights discourse.

Index

Routledge Global Institutions Series

70 Feminist Strategies in International Governance (2013)
edited by Gülay Caglar (Humboldt University, Berlin), Elisabeth Prügl (the Graduate Institute of International and Development Studies, Geneva), and Susanne Zwingel (the State University of New York, Potsdam)

69 The Migration Industry and the Commercialization of International Migration (2013)
edited by Thomas Gammeltoft-Hansen (Danish Institute for International Studies) and Ninna Nyberg Sørensen (Danish Institute for International Studies)

68 Integrating Africa (2013)
Decolonization's legacies, sovereignty, and the African Union
by Martin Welz (University of Konstanz)

67 Trade, Poverty, Development (2013)
Getting beyond the WTO's Doha deadlock
edited by Rorden Wilkinson (University of Manchester) and James Scott (University of Manchester)

66 The United Nations Industrial Development Organisation (2012)
Industrial solutions for a sustainable future
by Stephen Browne (FUNDS Project)

65 The Millennium Development Goals and Beyond (2012)
Global development after 2015
edited by Rorden Wilkinson (University of Manchester) and David Hulme (University of Manchester)

64 International Organizations as Self-Directed Actors (2012)
A framework for analysis
edited by Joel E. Oestreich (Drexel University)

63 Maritime Piracy (2012)
by Robert Haywood (One Earth Future) and Roberta Spivak (One Earth Future Foundation)

62 United Nations High Commissioner for Refugees (UNHCR) (2nd edition, 2012)
by Gil Loescher (University of Oxford), Alexander Betts (University of Oxford), and James Milner (University of Toronto)

61 International Law, International Relations, and Global Governance (2012)
by Charlotte Ku (University of Illinois)

60 Global Health Governance (2012)
by Sophie Harman (City University, London)

59 The Council of Europe (2012)
by Martyn Bond (University of London)

58 The Security Governance of Regional Organizations (2011)
edited by Emil J. Kirchner (University of Essex) and Roberto Domínguez (Suffolk University)

57 The United Nations Development Programme and System (2011)
by Stephen Browne (FUNDS Project)

56 The South Asian Association for Regional Cooperation (2011)
An emerging collaboration architecture
by Lawrence Sáez (University of London)

55 The UN Human Rights Council (2011)
by Bertrand G. Ramcharan (Geneva Graduate Institute of International and Development Studies)

54 The Responsibility to Protect (2011)
Cultural perspectives in the Global South
edited by Rama Mani (University of Oxford) and Thomas G. Weiss (The CUNY Graduate Center)

53 The International Trade Centre (2011)
Promoting exports for development
by Stephen Browne (FUNDS Project) and Sam Laird (University of Nottingham)

52 The Idea of World Government (2011)
From ancient times to the twenty-first century
by James A. Yunker (Western Illinois University)

51 Humanitarianism Contested (2011)
Where angels fear to tread
by Michael Barnett (George Washington University) and Thomas G. Weiss (The CUNY Graduate Center)

50 The Organization of American States (2011)
Global governance away from the media
by Monica Herz (Catholic University, Rio de Janeiro)

49 Non-Governmental Organizations in World Politics (2011)
The construction of global governance
by Peter Willetts (City University, London)

48 The Forum on China-Africa Cooperation (FOCAC) (2011)
by Ian Taylor (University of St. Andrews)

47 Global Think Tanks (2011)
Policy networks and governance
by James G. McGann (University of Pennsylvania) with Richard Sabatini

46 United Nations Educational, Scientific and Cultural Organization (UNESCO) (2011)
Creating norms for a complex world
by J.P. Singh (Georgetown University)

45 The International Labour Organization (2011)
Coming in from the cold
by Steve Hughes (Newcastle University) and Nigel Haworth (University of Auckland)

44 Global Poverty (2010)
How global governance is failing the poor
by David Hulme (University of Manchester)

43 Global Governance, Poverty, and Inequality (2010)
edited by Jennifer Clapp (University of Waterloo) and Rorden Wilkinson (University of Manchester)

42 Multilateral Counter-Terrorism (2010)
The global politics of cooperation and contestation
by Peter Romaniuk (John Jay College of Criminal Justice, CUNY)

41 Governing Climate Change (2010)
by Peter Newell (University of East Anglia) and Harriet A. Bulkeley (Durham University)

40 The UN Secretary-General and Secretariat (2nd edition, 2010)
by Leon Gordenker (Princeton University)

39 Preventive Human Rights Strategies (2010)
by Bertrand G. Ramcharan (Geneva Graduate Institute of International and Development Studies)

38 African Economic Institutions (2010)
by Kwame Akonor (Seton Hall University)

37 Global Institutions and the HIV/AIDS Epidemic (2010)
Responding to an international crisis
by Franklyn Lisk (University of Warwick)

36 Regional Security (2010)
The capacity of international organizations
by Rodrigo Tavares (United Nations University)

35 The Organisation for Economic Co-operation and Development (2009)
by Richard Woodward (University of Hull)

34 Transnational Organized Crime (2009)
by Frank Madsen (University of Cambridge)

33 The United Nations and Human Rights (2nd edition, 2009)
A guide for a new era
by Julie A. Mertus (American University)

32 The International Organization for Standardization (2009)
Global governance through voluntary consensus
by Craig N. Murphy (Wellesley College) and JoAnne Yates (Massachusetts Institute of Technology)

31 Shaping the Humanitarian World (2009)
by Peter Walker (Tufts University) and Daniel G. Maxwell (Tufts University)

30 Global Food and Agricultural Institutions (2009)
by John Shaw

29 Institutions of the Global South (2009)
by Jacqueline Anne Braveboy-Wagner (City College of New York, CUNY)

28 International Judicial Institutions (2009)
The architecture of international justice at home and abroad
by Richard J. Goldstone (Retired Justice of the Constitutional Court of South Africa) and Adam M. Smith (Harvard University)

27 The International Olympic Committee (2009)
The governance of the Olympic system
by Jean-Loup Chappelet (IDHEAP Swiss Graduate School of Public Administration) and Brenda Kübler-Mabbott

26 The World Health Organization (2009)
by Kelley Lee (London School of Hygiene and Tropical Medicine)

25 Internet Governance (2009)
The new frontier of global institutions
by John Mathiason (Syracuse University)

24 Institutions of the Asia-Pacific (2009)
ASEAN, APEC, and beyond
by Mark Beeson (University of Birmingham)

23 United Nations High Commissioner for Refugees (UNHCR) (2008)
The politics and practice of refugee protection into the twenty-first century
by Gil Loescher (University of Oxford), Alexander Betts (University of Oxford), and James Milner (University of Toronto)

22 Contemporary Human Rights Ideas (2008)
by Bertrand G. Ramcharan (Geneva Graduate Institute of International and Development Studies)

21 The World Bank (2008)
From reconstruction to development to equity
by Katherine Marshall (Georgetown University)

20 The European Union (2008)
by Clive Archer (Manchester Metropolitan University)

19 The African Union (2008)
Challenges of globalization, security, and governance
by Samuel M. Makinda (Murdoch University) and F. Wafula Okumu (McMaster University)

18 Commonwealth (2008)
Inter- and non-state contributions to global governance
by Timothy M. Shaw (Royal Roads University)

17 **The World Trade Organization (2007)**
Law, economics, and politics
by Bernard M. Hoekman (World Bank) and Petros C. Mavroidis (Columbia University)

16 **A Crisis of Global Institutions? (2007)**
Multilateralism and international security
by Edward Newman (University of Birmingham)

15 **UN Conference on Trade and Development (2007)**
by Ian Taylor (University of St. Andrews) and Karen Smith (University of Stellenbosch)

14 **The Organization for Security and Co-operation in Europe (2007)**
by David J. Galbreath (University of Aberdeen)

13 **The International Committee of the Red Cross (2007)**
A neutral humanitarian actor
by David P. Forsythe (University of Nebraska) and Barbara Ann Rieffer-Flanagan (Central Washington University)

12 **The World Economic Forum (2007)**
A multi-stakeholder approach to global governance
by Geoffrey Allen Pigman (Bennington College)

11 **The Group of 7/8 (2007)**
by Hugo Dobson (University of Sheffield)

10 **The International Monetary Fund (2007)**
Politics of conditional lending
by James Raymond Vreeland (Georgetown University)

9 **The North Atlantic Treaty Organization (2007)**
The enduring alliance
by Julian Lindley-French (Center for Applied Policy, University of Munich)

8 **The World Intellectual Property Organization (2006)**
Resurgence and the development agenda
by Chris May (University of the West of England)

7 **The UN Security Council (2006)**
Practice and promise
by Edward C. Luck (Columbia University)

6 **Global Environmental Institutions (2006)**
by Elizabeth R. DeSombre (Wellesley College)

5 **Internal Displacement (2006)**
Conceptualization and its consequences
by Thomas G. Weiss (The CUNY Graduate Center) and David A. Korn

4 **The UN General Assembly (2005)**
by M. J. Peterson (University of Massachusetts, Amherst)

3 **United Nations Global Conferences (2005)**
by Michael G. Schechter (Michigan State University)

2 **The UN Secretary-General and Secretariat (2005)**
by Leon Gordenker (Princeton University)

1 **The United Nations and Human Rights (2005)**
A guide for a new era
by Julie A. Mertus (American University)

Books currently under contract include:

The Regional Development Banks
Lending with a regional flavor
by Jonathan R. Strand (University of Nevada)

Millennium Development Goals (MDGs)
For a people-centered development agenda?
by Sakiko Fukada-Parr (The New School)

Peacebuilding
From concept to commission
by Robert Jenkins (The CUNY Graduate Center)

UNICEF
by Richard Jolly (University of Sussex)

The Bank for International Settlements
The politics of global financial supervision in the age of high finance
by Kevin Ozgercin (SUNY College at Old Westbury)

International Migration
by Khalid Koser (Geneva Centre for Security Policy)

Human Development
by Richard Ponzio

Religious Institutions and Global Politics
by Katherine Marshall (Georgetown University)

The Group of Twenty (G20)
by Andrew F. Cooper (Centre for International Governance Innovation, Ontario) and Ramesh Thakur (Balsillie School of International Affairs, Ontario)

The International Monetary Fund (2nd edition)
Politics of conditional lending
by James Raymond Vreeland (Georgetown University)

The UN Global Compact
by Catia Gregoratti (Lund University)

Institutions for Women's Rights
by Charlotte Patton (York College, CUNY) and Carolyn Stephenson (University of Hawaii)

International Aid
by Paul Mosley (University of Sheffield)

Global Consumer Policy
by Karsten Ronit (University of Copenhagen)

The Changing Political Map of Global Governance
by Anthony Payne (University of Sheffield) and Stephen Robert Buzdugan (Manchester Metropolitan University)

Coping with Nuclear Weapons
by W. Pal Sidhu

Global Sustainability
by Tapio Kanninen

Private Foundations and Development Partnerships
by Michael Moran (Swinburne University of Technology)

The International Politics of Human Rights
edited by Monica Serrano (Colegio de Mexico) and Thomas G. Weiss (The CUNY Graduate Center)

For further information regarding the series, please contact:

Craig Fowlie, Publisher, Politics & International Studies
Taylor & Francis
2 Park Square, Milton Park, Abingdon
Oxford OX14 4RN, UK
+44 (0)207 842 2057 Tel
+44 (0)207 842 2302 Fax
Craig.Fowlie@tandf.co.uk
www.routledge.com